AF575914

JH
AF
71 309

AVIATION

Vought A-7 Corsair II

The US Navy and US Air Force's Light Attack Aircraft

MAT "IRISH" GARRETSON

Library of Congress Control Number: 2020952586

Front cover photo courtesy of John "Lites" Leenhouts
Designed by Justin Watkinson
Type set in Impact/Minion Pro/Univers LT Std

ISBN: 978-0-7643-6261-3
Printed in China

Published by Schiffer Publishing, Ltd.
4880 Lower Valley Road
Atglen, PA 19310
Phone: (610) 593-1777; Fax: (610) 593-2002
E-mail: Info@schifferbooks.com
www.schifferbooks.com

For our complete selection of fine books on this and related subjects, please visit our website at www.schifferbooks.com. You may also write for a free catalog.

Acknowledgments

I am deeply indebted to countless individuals who have assisted me in the creation of this book. Many of the photos you will find between these pages have never before appeared in print. For that, I am especially thankful to David F. Brown, Tom "Burner" Brown, Charlie "Ozark" Cook, Dave "Hide" Dollarhide, Julio Fuentes, Jeff "Spock" Greer, Michael Grove, Mic Hamilton, Jeffery "Sundance" Harrison, Jan Jacobs, Rodd Karp, John "Lites" Leenhouts, Rick Morgan, James "Jimmy Jet" O'Hora, Apostolous "Papilos" Papadopoulos, Richard "Smokey" Powell, Denny Sapp, Dave "Spidey" Schneiderman, Keith Swendsen, Bob "Boo" Thomas, Bill Thomas, and Ric Versteeg. Thank you, gentlemen, for your generosity.

The staff and volunteers at the National Naval Aviation Museum provided me with unrestricted access to their archives, as well as a workspace during my many visits. Thank you, Sterling Gilliam, Buddy Macon, Hill Goodspeed, Marc Levitt, Frank Turchi, Pam Thomas, and Jared Galloway for the support and hospitality.

My fellow bubbas at the A-7 Corsair II Association have been incredibly supportive of this effort, some of whom have already been mentioned. I am proud to work alongside both those I've mentioned and those I am about to. Many thanks to Mike "Foot" Wilson, Charlie "Chaz" Maynard, Bill "BJ" Minkoff, Jeff "Jeep" Stivers, Ray "Wadds" Waddell, Sheree "Wonder Woman" Wolf, John "Rat" Leslie, Carl "Tank" Tankersley, William "Shoes" Johnson, Tim Palmer, Bob "Dirtball" Ryan, Jim Larkin, Mike "Magwai" White, and George "Spider" Webb. I value your friendships.

A special thanks goes to Mike "Lobster" Fitzgerald for his support and generosity in writing the foreword to this book.

A few other friends, bubbas, and mentors also deserve acknowledgment, both for their enthusiasm for this book and, for many, their lifelong support of my A-7 efforts. They include Kevin "Hoser" Miller, Shand Gause, Dick Atkins (RIP), Sol Love, Gerry "Ho Ho" Hoewing, Jerry "Arrow" Palmer, Mike "Carlos" Johnson, Garland Hogan, Barry Hendrix, Kyle Kirby, Lenore Taylor, and Dewey Larson.

Finally, I would like to assume that, in picking up this book, you share my passion for the A-7 Corsair II. If I am correct in this assumption, please accept my personal invitation to join us in the A-7 Corsair II Association. We are a nonprofit organization dedicated to preserving the legacy of this great plane, along with the men and women who flew and maintained her. Please go to www.corsair2.us to learn more.

Dedication

This book is dedicated to my family. Amie, Jack, and Thomas have been incredibly supportive of my passion for all things A-7. That passion is nothing when compared to the love I have for them.

Contents

Foreword

The history of naval aviation is replete with new concepts both evolutionary and revolutionary. Initially, carrier-based aircraft evolved from balloons and crow's nests to become scouting platforms for the main battleships of the fleet. They became revolutionary with Pearl Harbor and the obsolescence of the battleship during World War II. Aircraft evolved, becoming faster and more lethal, and the evolutionary advent of the jet age changed the dynamics of dogfighting and weapons delivery. The revolution came with air-to-air missiles and guided bombs.

The A-7 has a unique place in naval aviation, since it is both an evolutionary aircraft and, with the introduction of the A-7E, a revolutionary weapons system. Born of the need during the Vietnam War for an attack aircraft to increase payload from the A-4 Skyhawk, Vought quickly assembled parts from its F-8 Crusader fighter to build the Corsair II. The airframe itself was underpowered, and not too maneuverable, but it could carry a load of bombs to the target. With the postwar introduction of the A-7E (and A-7D in the Air Force), a new breed of aircraft emerged from which every subsequent aircraft improved upon.

The heads-up display (HUD), inertial navigation system, integrated navigation/bombing weapons system, and digital flight controls truly ushered in the digital age of combat aircraft. During my twenty years flying the A-7, the airframe remained mostly as designed, but the addition of missiles such as High Speed Anti Radiation (HARM); Stand-off, precision-guided Land Attack Missile (SLAM); and Extended Range Data Link (ERDL) Walleye, along with mines and conventional ordnance that was computer-delivered, made it a lethal platform that paced the threats it had to face. The Corsair II earned its way into the first strike on Baghdad in Desert Storm due to its unmatched capability to deliver HARM missiles and its electronic suite of warning receiver, chaff, and self-protection jammer. The A-7E of 1991 had little resemblance to the capability of the 1970 A-7A. It was truly revolutionary.

Mat Garretson has lived through this metamorphic life of the Corsair II and has spent several years chronicling the people and aircraft that made it a mainstay of naval aviation, as well as an important part of the Air Force attack community. His leadership as executive director in the Corsair II Association has kept the dream of flying the "Harley of the Fleet" alive in many of his fellow pilots who loved flying the Corsair. The legend of the Corsair is still alive in the hearts and minds of those who slipped the surly bonds of earth in this magnificent machine.

VAdm. Mark "Lobster" Fitzgerald, USN (Ret.)

Prologue

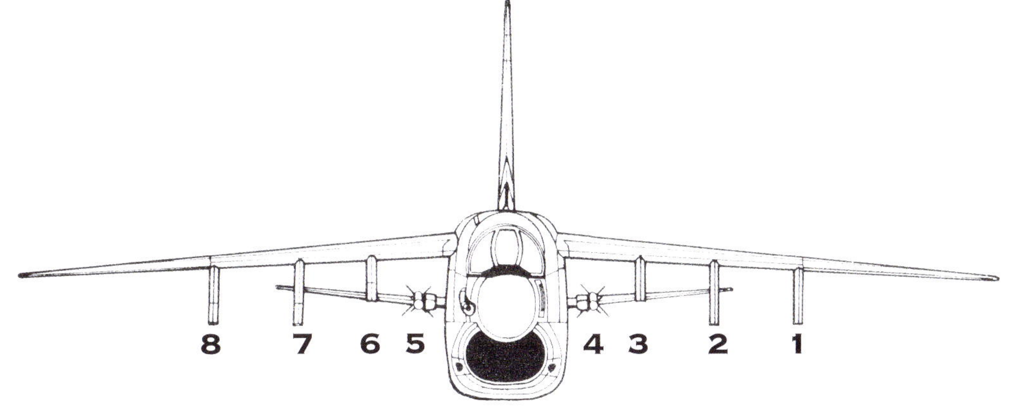

A-7 WEAPONS STATIONS

"Today we are seeing the roll-out and first public flight of an aircraft of unusual significance. This event represents a notable achievement on the part of the aircraft industry team represented by Ling-Temco-Vought, Pratt & Whitney, and the many subcontractors involved in this program. This talented team designed, built, and flown [sic] this impressive machine twenty-six days ahead of a most-demanding schedule. They have accumulated thirty-nine successful flights and 38.2 valuable flight hours before this first public showing. We look forward with enthusiasm to seeing the Corsair II in flight just a few minutes from now.

In recent years the nuclear superiority of the United States has caused the communists to pursue their imperialistic aims by manipulation of force involving limited means. The war they have created in Vietnam is typical. The communists have explicitly warned the world that this is to be their pattern for the future.

The A-7 is a tangible and direct response to that challenge. It is, in addition, a demonstration of our decision-making capacity, and our industrial capability.

A bare thirty months ago we had not even begun design competition for this airplane. We went from the first announcement of the limited design competition to letting the contract in a scant ten months. That is less than half the normal time. We committed ourselves—and now twenty months later, the team headed by Ling-Temco-Vought is making good on our commitment."

Paul H. Nitze
Secretary of the Navy
A-7 Roll-Out Ceremony
Dallas, November 2, 1965

Author's note

Over the lifetime of the A-7, the company who built them went through a number of name changes … from Chance Vought to Ling-Temco-Vought, then LTV, Vought Aerospace, and, eventually, Vought Corporation. For simplicity's sake, when describing the company, "Vought" will be employed.

There were eight weapons stations on the A-7, three on each of the wings, and one on each side of the fuselage. In describing ordnance in any of this book's photos, station locations utilize the reference diagram presented below.

CHAPTER 1

Winning the VAL

In 1962, the United States Navy found itself in the market for a new light attack aircraft. Since 1956, the light attack role was being fulfilled by the Douglas A-4 Skyhawk. While a great aircraft in its own right, the Skyhawk had been designed with one role in mind: providing the means of delivering a single nuclear warhead on an enemy target. With a changing mission profile, the Navy realized it needed an aircraft with longer range that was able to carry a larger payload and deliver it with greater accuracy than Douglas's bantam bomber. In early 1963, the chief of naval operations released its Sea-Based Strike Study, spelling out the details for a new aircraft dubbed the VAL, Navy nomenclature for fixed-wing (V), attack (A), light (L).

With Robert McNamara serving as secretary of defense, procurement of this new attack plane would follow his strict rules of efficiencies. McNamara and company decreed that the VAL would be powered by a nonafterburning version of Pratt & Whitney's new TF-30 turbofan, the same engine specified for the Navy's new fighter (at the time assumed to be the F-111B, an aircraft eventually rejected for the role, a move that made way for the F-14 Tomcat). A relatively new development, a turbofan engine offered greater fuel efficiencies over other jet engines of the day. In addition, requests for proposal (RFPs) were extended only to contractors of preexisting Navy aircraft, with the mandate being that their designs be based upon their aircraft. Using this criterion, the DoD felt that the winning design would have lower development costs and benefit from commonality of parts with the existing design. As a result, in May 1963, RFPs were requested from Grumman (using the A-6 Intruder as its basis), North American (FJ1 Fury), Douglas (A-4 Skyhawk), and Vought (F-8 Crusader).

Of these four contractors, none were as hungry as Dallas-based Vought. With production of their F-8 slated to come to an end in 1965, and with no other major contract in the works, winning the VAL was imperative for the company's survival. Internally, Vought set up two teams: the "Blue Team," focused on creating the best design possible, and the "Purple Team," charged to think like the competition so as to provide insights as to what the Blue Team was up against. Externally, Vought spooled up an impressive marketing effort to woo the Navy to the Vought design. When it came to working (and working with) the Navy, Vought had few peers. By 1964, the company had been providing reliable and capable combat aircraft for the Navy for almost fifty years; most notably, the outstanding F4U Corsair. The ranks of the company's management team were a Who's Who of retired naval aviators, including Vought's then president, Paul Thayer, a former World War II Navy ace, and J. W. Lankford, who flew dive-bombers during the war and now headed up Vought's full-court press on the Washington brass.

Understanding the Navy as well as they did, Vought recognized the need for their VAL submission to bear as similar an appearance to their F-8 as possible. Over the course of the competition, it was increasingly understood by both the Navy and the competing contractors that while commonality with an existing design was desirable, the reality was that this new attack aircraft would be, for all intents and purposes, a completely new aircraft. Despite this, to ensure buy-in from the Navy, Vought held firm to designing a plane that looked like the F-8. In the end, while there was a striking resemblance to the Crusader, Vought's entry had little to no commonality to it.

The Vought proposal—internally known as V463—certainly bore an outward resemblance to the company's gunslinger, with a single chin intake, high-mounted anhedral wing, and low UHT (unit horizontal tail) with noted dihedral. Obvious differences between the V463 and F-8 included the fact that the former had a shorter (by nearly 8 feet), much-wider fuselage (both to incorporate greater internal fuel capacity and more avionics in the cockpit), as well as a beefier wing to accommodate the eight parent stations envisioned for the aircraft. Addressing the F-8's poor maintenance record, Vought emphasized ease of access to all major componentry in its VAL entry, providing a number of waist-high panels, as well as a (relatively) easy, modular means of removing the TF-30 engine.

In reviewing the proposals from the four competitors, the Navy found that two designs—Vought's and North American's—proved the most capable of fulfilling the specified mission requirements. When it came to acquisition cost, it was the Douglas and Vought designs that led the pack. Given these two realities, the Vought entry clearly provided the Navy the best "bang for the buck," and on February 11, 1964, the company was named the winner of the VAL competition, with the resulting aircraft being designated as the A-7. In early 1965, Vought settled on a name for their new airplane—the Corsair II.

Artist's concept of the Vought V463, the design that eventually led to the A-7A. *Author's collection*

While outwardly looking like any other Skyhawk, Douglas's VAL entry, the A4D-6, dwarfed its predecessor, as shown here in comparison of scale models. With a larger wing area and longer, wider empennage to incorporate the TF-30, the A4D-6 was an entirely new airplane, moving Douglas from the presumptive shoe-in for the VAL contract to one that was competing on a more level field with the other contractors. *Author's collection*

The Navy invited four manufacturers to submit proposals for their VAL competition. The RFP specified that all designs were to be based on an existing airframe. Grumman's entry was the G12. Essentially a single-seat A-6, stripped of most of its high-tech avionics, Grumman decided to retain the twin J-52 engine design of the A-6 instead of incorporating the DoD's desire to use a single TF-30 turbofan. This decision, along with the fact that the G12 was nearly twice as expensive as the competition's entries, doomed the concept from the outset. *Author's collection*

Vought technicians constructing the full-scale V463 mockup. Produced from wood and foam core, the mockup would help Vought both sell the aircraft to the Navy as well as serve as proof of concept for many design features. Note the pointed aspect of the radome, which was eventually blunted, primarily as a means to reduce the overall length of the aircraft. *Author's collection*

The finished mockup on display in a Vought warehouse. The mockup featured only the port side wing, which necessitated the use of a four-point rig mounted from the wing to the warehouse ceiling. Note that the radome is still pointed. Immediately aft of the nose is a small, four-gill cooling vent for the radar, later redesigned to a larger, three-gilled vent. Also of note is the F-8-style, two-strut boarding stirrup. A-7As and Bs utilized a single-strut stirrup. *Author's collection*

This rear quarter view of the mockup showcases the single-piece tailpipe fairing and belly panels removed so as to easily access or replace the TF-30. Note that the underside of the tailpipe fairing features a notch to accommodate the engine tailpipe—a feature deemed unnecessary in production aircraft. Note also the lower forward sweep of the rudder, something borrowed from the F-8's design but not incorporated in actual A-7s. *Author's collection*

Here a Vought technician can be seen in front of the mockup's portside avionics bay. Unique for its day, the avionics bay provided waist-high access to critical components. To the left of this bay is the access to the pilot's liquid oxygen (LOX) canister, and to the left of it a large access panel to the Colt Mk. 12 20 mm cannon. This gun access door was substantially redesigned and decreased in size in the actual aircraft. *Author's collection*

Same view, with all of the panels "buttoned up." Notice the trailing edge of the vertical stab's fin cap. Vought executive Jessie Santamaria is credited with the forward sweep to the trailing edge. This "notch" to the tail was done as a last-minute change to the submitted design, made in an effort to reduce the A-7's footprint on the carrier. While the addition of a radar-warning receiver fairing to the fin cap negated the space savings, this distinctive design element was incorporated in all variants of the aircraft with the exception of the YA-7F. *Author's collection*

CHAPTER 2

Prototypes and Tests

In June 1964, having named Vought the winner of the VAL competition, the Navy issued the company an initial contract for seven A-7A test articles and thirty-five production aircraft. A thorough inspection and review of the Corsair II mockup resulted in 154 recommended changes, all of which were incorporated into the prototype. On September 27, 1965—just eighteen short months after Vought was named winner of the VAL—the first A-7A prototype first took to the skies. Vought's chief test pilot, John W. Konrad, had the honor of this first flight, and he would later state that the A-7A exhibited very similar flight characteristics to the North American F-86 Sabre.

Over the next year, the seven prototype aircraft were put through their paces, providing data that would improve follow-on, production aircraft. While these airframes were utilized for a variety of tests, each had a primary purpose; namely:

#1	BuNo 152580	Prototype
#2	BuNo 152581	Spin
#3	BuNo 152582	Structural
#4	BuNo 152648	Weapons System Testing
#5	BuNo 152649	Avionics
#6	BuNo 152650	Armament
#7	BuNo 152651	Carrier Suitability

Testing confirmed that the Corsair II would indeed live up to expectations, either meeting or exceeding contracted specifications. In short, the Navy was getting ready to field an aircraft that could carry twice the load of an A-4, could fly farther, was more accurate, and was easier to maintain. The primary gripe from most pilots centered on engine performance. The A-7A utilized the Pratt & Whitney TF-30-P-6, a turbofan that provided 11,350 pounds of thrust. The follow-on variant, the A-7B, featured a TF-30-P408 powerp lant that offered up 12,200 pounds. Eventually, the A-7E would feature an engine that provided nearly 15,000 pounds of thrust—better, but it was never enough for the Corsair II to shake off the oft-repeated comment "The A-7 ain't fast—but it sure is slow!"

The A-7 finally earned its sea legs in November 1966, during carrier suitability trials off the Virginia Capes. Three pilots from the Naval Test Center in Patuxent River conducted these BIS (Board of Inspection and Survey) tests aboard the Navy's newest carrier, the USS *America*. A-7As #7 (BuNo 152651) and #15 (BuNo 152658) were utilized for the tests. A total of eighty-nine touch-and-gos, seventy-three arrested landings, and seventy-five catapult shots were made during the weeklong trial. This and subsequent carrier suitability tests revealed the potential for engine compressor stalls due to the ingestion of catapult steam. This potential resulted in limitations to the A-7A's gross takeoff weight and served to emphasize the need for a more powerful powerp lant.

While testing would continue, the initial results were so successful that A-7s started arriving to RAG (Replacement Air Group) squadrons on both coasts, within a year of the Corsair II's first flight—an achievement unheard of these days.

A-7A #1 starts to take shape. Note the "gull wing" doors immediately behind the canopy. These doors accessed the canisters of 20 mm ammunition for the two Colt Mk. 12 cannons. *Author's collection*

Nearly complete, A-7A #1 leaves for the painting shed. Note that the ejection seat has yet to be installed, and given how high the aircraft sits on its main wheels, it would appear that the engine hasn't been installed yet, either. *Author's collection*

Vought's chief test pilot, John W. Konrad, prepares for the A-7A's first flight, September 27, 1965. In later years, Konrad would compare the performance and handling of the A-7A to that of an F-86 Sabre. *Author's collection*

A-7A #1 at its public debut and flight, November 2, 1965. Now sporting Navy livery (and BuNo 152580), Vought threw a big shindig that day, which included a Texas-style barbecue for roughly 1,000 visiting dignitaries and guests. *Author's collection*

The November 2 debut featured three A-7As. A fully loaded Alpha took to the skies first, showcasing the type's weapons delivery capabilities. This was followed by the launch of a clean aircraft, which highlighted the A-7A's performance. While these two Alphas performed overhead, the third remained in front of the viewing stand, where Vought technicians performed an engine removal in less than twenty-five minutes. *Author's collection*

A-7A #1 taking to the skies November 2, 1965. Sadly, this aircraft would be lost nearly five months later during testing at China Lake. *Author's collection*

While bearing a family resemblance to the F-8 Crusader, the A-7 was a completely new aircraft. Here Prototype #1 flies in formation with an F-8 over the Texas countryside sometime in 1965. *Naval Aviation Museum*

Prototype #2, BuNo 152581, carries a TER of three dummy Mk. 82s on station #8, and six more on a MER on station #7. Similar in paint scheme to A-7A prototype #1, this aircraft sports a gray radome. Note as well the positioning of the BuNo, below the horizontal tail. Production models would see it repositioned below the "Navy" stencil. *David F. Brown collection*

A-7A #15 (BuNo 158658) approaches the ramp of the USS *America* during the BIS trials, November 1966.
Bob Lawson Collection, National Naval Aviation Museum

A-7A #7 (BuNo 152651) taxies toward the catapult shuttle in preparation for launch during BIS trials, while A-7A #15 waits its turn behind the jet blast deflector. *Author's collection*

Catapult steam ingestion was an ongoing issue with the A-7A, resulting in a limitation on gross launch weight. This issue was resolved with the introduction of A-7B, with its more powerful engine. *Author's collection*

Capt. Don Engen (*far left*), skipper of the USS *America,* chats with Pax River test pilots (*left to right*) Cmdr. Don Lyman, Lt. Robert Coffman, and LCmdr. Fred Hueber during the weeklong BIS carrier suitability trials, November 1965. Huber had the honor of being the first pilot to both land and launch an A-7 from an aircraft carrier. *Author's collection*

A-7A #3 (BuNo 152582) loaded down with an impressive number of inert Mk. 81 bombs. The A-7 could delivery nearly every weapon in the Navy's arsenal. *Author's collection*

Another shot of A-7A #3, providing a great view of the aircraft's underside. As originally proposed, the A-7 would have featured four parent stations on each wing. Studies concluded that a fourth station would have provided no real additional ability to carry more ordnance, since the station closest to the fuselage would have been severely limited, with most loads interfering with the main landing-gear door. *Naval Aviation Museum*

The twelfth A-7A built, BuNo 152656, taxies out from the Vought plant for its first flight. Interesting to note is the lack of a refueling probe fairing, a trait shared by all early A-7As. This was eventually addressed. *Author's collection*

A close-up view of the refueling probe on A-7A #1. Original accommodation for the probe in stowed position was a shallow well for the probe, and a small 'bumper" on the fuselage. Excessive wind noise necessitated a faring around both the well and the probe shaft. *Author's collection*

A-7A #7 (BuNo 152651) traps aboard the USS *Independence* in February 1967, during carrier qualifications of the A-7 with operational loads. This Alpha is loaded with a 300-gallon Aero 1D fuel tank on the inboard station, with an inert Mk. 82 200-lb. general-purpose bomb. *Author's collection*

CHAPTER 3

Introduction to the Fleet

Even before the first A-7's carrier suitability test was conducted, A-7As were on their way to Navy squadrons. The first of these deliveries took place on October 14, 1966, with two Alphas departing the Vought factory in Texas, bound for NAS Cecil Field, Florida. Upon arrival, they reported to the "Hell Razors" of VA-174, the A-7 East Coast RAG (Replacement Air Group), or training squadron. The following month saw a similar delivery of two A-7As, this time to Lemoore, California, which were assigned to the "Flying Eagles" of VA-122, 174's West Coast equivalent. Both squadrons would quickly acquire additional aircraft and experience and would soon be training their personnel and pilots as the A-7A prepared to make its way to the fleet.

The "Sidewinders" of VA-86 were the first fleet A-7A squadron to stand up. Home-ported at NAS Cecil Field, VA-86 received their Alphas in February 1967. By June 1967, the Sidewinders' sister squadron, the "Marauders" of VA-82, would also receive their A-7As. That same month saw the first Alphas delivered to a West Coast fleet squadron, the "Argonauts" of VA-147. By the end of 1967, three additional A-7A squadrons would join the ranks: VA-37 "Bulls" and VA-105 "Gunslingers" at Cecil Field, and the "Warhawks" of VA-97 at Lemoore. In January 1968, the Lemoore-based "Royal Maces" of VA-27 would get their Alphas too, rounding out the initial roster of seven fleet squadrons to use the A-7A. Each squadron quickly busied themselves in learning how to employ and maintain the A-7A as a weapon. With the war in Vietnam escalating, the learning curve for all squadrons was steep, quick paced, and deadly serious.

By the middle of 1968, the A-7B variant made its way to the fleet, with both Cecil-based VA-87 and VA-146 at Lemoore getting theirs in June of that year. By 1970, one RAG and eleven more fleet squadrons would be outfitted with A-7Bs, bringing the total number of A-7 squadrons to twenty-five. The Bravo was essentially an A-7A with an upgraded engine—the TF-30-P-408—initially providing an additional 845 pounds of thrust, with future improvements providing another 1,200 pounds. The P-408 helped eliminate the catapult steam ingestion issues, but most pilots still felt that the A-7 could benefit from a more powerful powerp lant.

All in all, the A-7 did a great job of meeting or exceeding the Navy's expectations. In the A-7 they got a great bomber, able to carry more—and carry it a greater distance—than its predecessor. And it did all of this at a relatively low fixed per-unit cost. The arrival of the A-7 was timely, since the United States was stepping up its commitment to South Vietnam. The need for a better CAS (close air support) aircraft was such that the A-7 would be conducting combat sorties over Southeast Asia within twenty-five months of its first flight.

VA-174's skipper, Cmdr. Don Ross, preparing to depart the Vought factory with the squadron's first A-7A, BuNo 152649, on October 14, 1966. *Author's collection*

The A-7A featured the Texas Instruments' AN/APQ-116, which was supposedly a true terrain following radar (TFR). The company provided A-7 pilots with these patches to tout the radar's capability. In recalling the -116, naval aviator Tom Brown said, "… most of us never trusted TFR or took it seriously. I used it only in VFR conditions, and it was nowhere near ready for prime time." *Author's collection*

VA-174 XO (Executive Officer) Cmdr. Richard Vaillancourt flies wing on Cmdr. Don Ross as the two pass over NAS Cecil Field. Vaillancourt would soon move, serving as the first commanding officer of VA-97, the second West Coast fleet squadron. *Bob Lawson Collection, National Naval Aviation Museum*

October 14, 1966
VA-174 skipper Ross hands over the logbooks for the first two A-7As to RAdm. H. H. Caldwell, commander, Fleet Air, Jacksonville. Between the two are (*left to right*) VAdm. C. T. Booth, commander, Naval Air Forces, Atlantic Fleet; Capt. Jack Christiansen, CO of Carrier Air Wing 4; and Paul Thayer, president of LTV Aerospace. *Author's collection*

November 1966, the first A-7A assigned to VA-122 departed the Vought facility, bound for Lemoore, California. LCmdr. Lou Taylor was the pilot. The "Flying Eagles" of VA-122 served as the West Coast A-7 RAG or training squadron, while VA-174 was served as the East Coast A-7 RAG. *Author's collection*

Eagle 202 and 201 on an early flight over NAS Lemoore. *Naval Aviation Museum*

Within a year VA-122 made changes to their markings, forgoing the large yellow stripe on the tail for a yellow and gray trim immediately below the fin cap. Here, Eagle 216 (A-7A BuNo 153145) heads out to the range for some target practice with live, 500-pound Snakeyes. *Author's collection*

This photo, taken January 1968, shows Eagle 201 decorated with colorful squadron markings, and the moniker "Corsair College." Interesting to note the squadron designation on the tail, VA(F)-122. It would take thirty-one years before a 122 aircraft truly wore a "VFA" designation. *Naval Aviation Museum*

The "Sidewinders" of VA-86 received their Alphas in February 1967. This is Winder 414 (A-7A BuNo 152669), and the photo was taken at an airs how in August 1967; photographer and location are unknown. *National Naval Aviation Museum*

Four months after their sister squadron (VA-86) got theirs, the "Marauders" took delivery of their Alphas. The squadron's CAG bird (A-7A BuNo 153230) flies above the cloud deck, with an empty MER on station #8 and a live AIM-9 on its fuselage station. After two combat tours with VA-82, this aircraft would be reassigned to VA-105. On June 17, 1972, it would be lost to enemy fire near Ha Tinh, with her pilot, Lt. Larry Kilpatrick, being KIA. *Photo by Tom Brown*

A showroom-new Alpha (A-7A BuNo 153199) of VA-147 on the ramp at Lemoore. Before embarking on their first combat cruise, the "Argonauts" would receive upgraded A-7As that featured improved performance and survivability. The easiest way to identify the "older" Alphas is by the lack of an ECM antenna fairing above the rudder. *National Naval Aviation Museum*

Less than three weeks after receiving their A-7As, VA-147 went to the boat. The squadron's first-ever carrier qualifications took place aboard the USS *Ranger*, June 19–23, 1967. Five months later, the "Argonauts" would be back on *Ranger*, this time to conduct combat operations over North Vietnam. *Bob Lawson Collection, National Naval Aviation Museum*

VA-37 received their A-7As in August 1967. Here, Bull 310, BuNo 153268, sits on the ramp at Cecil, along with three other squadron Alphas. 310 is configured to provide inflight refueling services. On station #1 is a Douglas D701 refueling pod containing the hose and drogue, or "basket." An Aero 1D 300-gallon fuel tank is loaded on the right wing, station #8. Clearly visible is the addition of the fairing above the rudder. This fairing housed two ECM antennas, the AN/APR-25 and AN/ALQ-51A, and necessitated the need to relocate the tail formation light further forward. *Photo by Hideki Nagakubo, National Naval Aviation Museum*

Canyon Passage 406 (A-7A BuNo 153190) on the ramp. VA-105 received their Alphas in November 1967, three months after their sister squadron, VA-37, received theirs. Thirteen months later, both the "Gunslingers" and "Bulls" would be embarked on the USS *Kitty Hawk*, flying sorties from Yankee Station. *National Naval Aviation Museum*

The "Warhawks" of VA-97 were first equipped with early versions of the A-7A in October 1967. Here a two-ship of the early Alphas heads to the ranges at Fallon in preparation for their initial combat cruise, which departed in May of the following year. *Author's collection*

Sister squadron to VA-97 were the "Royal Maces" of VA-27. They received their Alphas in January 1968. This photo, taken in March 1969, shows Charger 613 (A-7A BuNo 153176) on the ramp at NAS Lemoore, having returned from their first combat deployment aboard *Constellation* on January 31 of that year. *National Naval Aviation Museum*

CHAPTER 4

The "Big D"

Once Vought was awarded the A-7 contract, the argument of "Why not a USAF variant?" became one increasingly heard in the halls of the Pentagon. At the time, the Air Force was internally discussing the needs for a new close air support (CAS) aircraft to replace the venerable F-100 Super Sabres, A-1 Skyraiders, and A-26 Invaders then tasked in that role. If these aging airframes didn't provide enough motivation to find a replacement, the Army's plans for a new attack helicopter certainly did. In this new helicopter, the Air Force saw a potential threat of being relieved of the CAS mission altogether.

In 1964, conventional thinking within the Air Force (bolstered by an internal study) held that the CAS mission could best be fulfilled by a mix of F-5s and stripped-down F-4s. Holding to a mantra of "faster is always better," the Air Force steadfastly believed that the F-5 was the better CAS choice because its supersonic performance made it infinitely more survivable than the subsonic A-7. This belief was held despite the fact that, at best, an F-5 could carry just one-fifth the payload of the A-7. At this point Robert McNamara personally entered the fray, instructing the Air Force to conduct a new study of the CAS need, limiting the USAF's options to aircraft already in production. McNamara further stipulated that this study was to assume that air superiority over the battlefield was a "given." In addition to the F-5 and F-4, the secretary specifically called out the A-7 and A-6 as aircraft the Air Force should consider. The resulting study concluded that the clear choice was the F-4, given that aircraft's speed and air-to-air capability, and the belief that compared to the A-7, the F-4 was more survivable in the CAS role. In short, the Air Force was unmoved from its predilection for speed and remained convinced that the slower A-7 would suffer unbearable attrition rates in a real-world environment.

Within months of the release of the requested study, external forces came to bear on the Air Force's way of thinking. To begin with, the DoD seriously doubted that Douglas had the capacity to produce a CAS variant of the F-4. The contractor's production line was already overburdened by the high demand for airframes requested by the Air Force, Navy, and Marines, not to mention their increasing export commitments. By the fall of 1965, the F-5 was conducting limited CAS sorties in Vietnam, and its performance was, to put it mildly, lackluster. The F-5's limited range and payload showed it to be a loser in the CAS role. Added to these realities was the fact that the Air Force was putting into motion its FX program for the next-generation air superiority fighter. This was a program the service dearly wanted, and one that would, in the end, result in the F-15 Eagle. Given these factors, and the DoD's stated preferences, it became clear to Air Force leadership that they were going to get the A-7.

By November 1965, the choice on the A-7 had been made (or, as many believed, made for them), and the Air Force wasted little time on providing inputs that would eventually improve not only its own variant but the Navy's as well. McNamara was fine on allowing the Air Force to refine the A-7, provided that the service take delivery of the first examples by January 1967. Initially, the USAF requested that their A-7 feature an afterburner, given that their fully loaded aircraft would not have the benefit of an aircraft carrier's catapult to compensate for the TF-30's marginal thrust. It was estimated that the addition of an afterburner on the TF-30 would result in an increased thrust of 15,000 pounds (compared to the 11,350 pounds being offered in the initial A-7A variant). In the end, the additional costs, degraded combat radius, and design delays to incorporate an afterburner into the A-7 proved to be a dead end. An alternative was eventually found in the Rolls-Royce Spey, a turbofan engine of superior performance to that of the TF-30, and already licensed for domestic production. Vought had already considered the Spey as a potential powerp lant for a proposed Canadian variant of the A-7,* so it was clear that the engine could fit into the airframe. Allison told the Air Force that they could deliver a nonafterburning version of the Spey that could provide 14,000 pounds of thrust, and do so within eighteen months. The green light was given, and this engine would eventually be known as the TF-41. When initial testing of this engine in an A-7 resulted in nearly 15,000 pounds of thrust, the Navy quickly became interested in utilizing the TF-41 for its new variant, the A-7E.

* In the end, the Canadians opted for the F-5, primarily on the basis of the assumption that this was the aircraft the USAF was going to use, and not the A-7. Had they known that the Air Force would eventually select the A-7, there is little doubt that Canada would have likewise done so.

In addition to an improved engine, the Air Force strongly felt that the A-7 needed an upgrade to its avionics suite; specifically, an ability to deliver its weapons with greater accuracy. As originally conceived, the A-7 was basically an A-4 that could carry a bigger payload farther. The A-7A relied on the same "iron bomb" delivery system of gunsight, pipper, and analog computer (and experienced pilot) for accurate delivery of its ordnance. As a result, accuracies of 20 to 40 Mils were considered normal. Experience then being earned in Vietnam had the Air Force convinced that it needed the A-7 to possess as accurate a delivery system as possible; 10 Mils or less. The Navy had also been studying the options in improved avionics that were becoming available via emerging technologies, but envisioned that these options would be installed on the aircraft that would eventually replace the A-7. The Air Force's insistence in a more accurate system resulted in the fast-tracking of getting state-of-the-art technology into the aircraft. In December 1966, the director of defense research and engineering within the DoD directed that the Air Force and Navy work together on a joint program to improve the A-7's avionics suite. This technology included the first-ever heads-up display (HUD), a digital weapons computer that interfaced with the radar, and a projected map display system (PMDS). Another mutual improvement to the aircraft made by the Air Force included removing the twin Mk. 12 20 mm cannons in favor of the M-61 Vulcan 20 mm rotary cannon.

While the Air Force may not have originally wanted the A-7, there is no doubting that their involvement in the program resulted in the A-7 being able to realize its full potential—at least earlier than the Navy would have done.

YA-7D #1 (s/n 67-14582) leaves the Vought production line, heading for the paint shop. February 7, 1968.
Author's collection

Vought chief test pilot John Konrad prepares for the first flight in the YA-7D, April 6, 1968. *Author's collection*

YA-7D #1 taxies out for its first flight. Note the USN-style refueling probe. The first sixteen A-7Ds to come off the assembly line featured this configuration, with subsequent airframes incorporating the dorsal refueling "hump" typical of USAF A-7s. *Author's collection*

The third YA-7D prototype (s/n 67-14585) made its first flight on September 26, 1968. Shortly thereafter it was shipped to Edwards AFB to undergo further testing. With its outboard wing panels removed and crated, it was loaded onto a Super Guppy and flown to Edwards. *Author's collection*

A great shot of the final configuration of the A-7D refueling receptacle, as seen from the boom operator's position on the refueling aircraft. Opening the door to the female-style plug on the A-7D was accomplished by the pilot pulling on a T handle immediately forward of the throttle. *Author's collection*

A-7A #7 (BuNo 152650) was used in the testing of a USAF-style refueling system. Note that the original configuration consisted of a "popup"=style receptacle, later replaced by a simpler, passive door and plug. Pilot for these tests was Vought's William Cato. This airframe later served as the gate guardian at NAS Cecil Field and is now on display at the Don Garlits' Racing Museum in Ocala, Florida. *Author's collection*

The cockpit of the A-7D featured many new, state-of-the-art innovations, including a heads-up display (top center) and Projected Map Display System (round gray screen, lower right). These two avionics on the A-7, combined with a digital navigation and weapons computer, ushered in an accuracy unequaled for its day. *Author's collection*

First A-7D to be delivered to a nontest squadron, s/n 68-8225, prepares to depart Vought for Luke AFB on September 1, 1969. Upon arrival it was assigned to the 4525th TWW, soon after designated 57th Wing. The man on the boarding ladder is Sol Love, head of the A-7 program at Vought. The pilot is Maj. C. W. McClarren, who also headed up the USAF contingency that deployed with VA-147 on the first-ever A-7 combat cruise. *Author's collection*

A trio of A-7Ds from the 57th Wing over the skies of Nevada. Based at Luke AFB, the 57th Wing would fine-tune the A-7D into a formidable weapon's platform. *Author's collection*

The 355th Tactical Fighter Wing (TFW) got the first of their eventual four squadrons of A-7Ds in 1972. Based at Davis Monthan AFB in Tucson, Arizona, the 355th would quickly deploy to Korat in Thailand to take part in Vietnam operations. This Big D (s/n 70-1032) eventually served in both the Michigan and Colorado ANGs before being retired. *Author's collection*

There's no doubting that the Air Force helped bring about improvements to the A-7 that would realize the aircraft's full potential. Here an A-7D (s/n 71-0293) flies wing on an A-7E (BuNo 157551) over the Arizona desert, sometime in 1972. *Author's collection*

CHAPTER 5

The A-7 at War

Over the course of its service with the US Navy and Air Force, the A-7 acquitted itself as a very reliable and capable warhorse. From iron bombs to antiradiation missiles, and from mines to smart bombs, the A-7 could deliver it—and deliver it with an accuracy unavailable before its arrival.

What follows is a photo essay of the A-7 in combat, 1967–1989.

Jason 311 (BuNo 153227) overflies the USS *Ranger*, returning from a combat mission, December 22, 1968. Note that the outboard wing stations have two Zuni rocket pod canisters (three empty, one full), and both fuselage stations have their AIM-9 Sidewinders. *Author's collection*

On November 4, 1967, the "Argonauts" departed the US on *Ranger*, destination Yankee Station. By early December, the squadron was conducting combat missions over Vietnam. Jason 307 (A-7A BuNo 153165) launches with a complement of sixteen Mk. 82 Snakeyes and two Sidewinders. Note that none of the MERs have bombs affixed to their inboard stations. *Naval Aviation Museum*

Jason 315 (BuNo 153219) prepares to trap aboard *Ranger*, sometime in January in 1969. Many of VA-147's combat-proven Alphas would find new life some twelve years later, refurbished at Vought, and sold to the Portuguese Air Force as A-7Ps. *Author's collection*

The A-7 would prove a great platform for Ironhand missions. Jason 317 (A-7A BuNo 153242) taxis forward for such a mission, loaded with four AGM-45 Shrikes. Note that the "VA-147" stenciled on the tail is in red, not black as it was originally applied. *Naval Aviation Museum*

Included on VA-147's roster for their first combat cruise was a contingency of four USAF pilots and twenty-one airmen, providing the service insight into how the A-7 performed in real life. Left to right: Maj. Gordon Williams, Maj. Chuck McClarren, Capt. Jim Alexander, and Capt. Nick Jones (maintenance officer). *Author's collection*

Returning to the boat after another mission, Jason 305 (A-7A BuNo 153222) carries back its Sidewinders and two full Zuni rocket pods. 305 would be lost late in the squadron's second combat cruise due to an engine flameout attributed to the failure of the accessory gear train. The pilot was forced to ditch but was quickly recovered. *Naval Aviation Museum*

A sticker Vought produced to commemorate VA-147's first combat cruise. *Author's collection*

VA-27 pilots flew their first combat sorties on May 28, 1968, while deployed on the USS *Constellation*. Here Mace 601 (A-7A BuNo 154344) flies trail on its wingman, heading back home after a combat sortie. Photo was taken with a NAMAR camera pod. *Photo courtesy of Rodd Karp*

As VA-147s prepared to depart Yankee Station, the USS *America* was heading that way and bringing two A-7As squadrons, VA-82 and VA-86, with it. Carrying identical loads of Snakeyes and Sidewinders, Streetcar 302 (A-7A BuNo 153218) flies wing on Winder 413 (A-7A BuNo 153263) head north on a mission. *Photo by Tom Brown*

Seven months after returning from their initial combat cruise, VA-97 was on its way back to Vietnam with CAG-14. Warace 313 (A-7A BuNo 154358) prepares to recover aboard *Constellation* after a sortie during this second deployment. On April 3, 1970, less than a month to go before the cruise ended, 313 was hit by a 37 mm shell. The pilot, Lt. H. P. Hoffman, continued on his run, dropped his bombs, and exited the target area. Sustained damage required Hoffman to eject near Da Nang, and he was rescued in short order. *Photo by Mic Hamilton*

Warace 302 (A-7A BuNo 153266) en route to the target, wearing almost as much zinc chromate as gull gray. The humid climate of Vietnam, combined with the salty air of carrier operations, kept every squadron's corrosion control teams very busy. *Photo by Bill Thomas.*

Mace 410 (A-7A BuNo 154352) drops its Snakeyes at altitude along with A-6s of VA-85. For its second combat cruise, the "Royal Maces" would utilize the 400 modex for their aircraft. After Vietnam, this Alpha would serve in the Naval Reserves, with VA-304, prior to being converted to an A-7P. It would be lost in a midair collision with another A-7P on July 1, 1985. *Photo courtesy of Rodd Karp*

VA-37's combat debut in A-7s would be aboard the *Kitty Hawk* as part of CVW-11. Along with their sister squadron, VA-105, the "Bulls" would be on cruise from December 30, 1968, until September 4, 1969. Here, Bull 312 (A-7A BuNo 153213) grabs the #3 wire, having returned from an Ironhand mission. VA-37 would operate the Alpha until the spring of 1973, during which time they'd make three cruises on *Saratoga,* one of which would be a Vietnam cruise. *Bob Lawson Collection, Naval Aviation Museum*

The A-7E would make its combat debut on May 26, 1970. The honor went to VA-146 in the opening days of an eight-month CVW-9 cruise that included sister squadron (and fellow Echo operator) VA-147. Diamond 310 (A-7E BuNo 157438) and 300 (A-7E BuNo 156810) cruise over the Tonkin Gulf on their way home to *America*. *Bob Lawson Collection, Naval Aviation Museum*

In September 1972, the 354th TFW was ordered to deploy the 353rd and 355th TFS from their home base in South Carolina to Korat AFB, Thailand. In all, seventy-two A-7Ds were deployed, where, on October 16, they became the first Big Ds to see combat. A-7D s/n 71-0354 is seen here, wearing standard SEA colors. *Author's collection*

Toward the end of CVW-9's 1970 combat cruise, it was joined by CVW-2 on *Ranger,* bringing with it VA-25 and VA-113, both having traded in their combat-tested Bravos for the new Echoes. Stingers 304 and 307 prepare to go feet dry, carrying with them Mk. 82s. *Author's collection*

A 354th TFW Big D taxis to the arming pits in preparation for another mission. Notice that there are no Mk. 82s mounted on the inboard stations of the MER. *Author's collection*

In March 1973, some of the 354th TFW's aircraft were reassigned to form the 3rd TFS, part of the 388th TFW, which was permanently based at Korat. Wearing the JH tailc ode, the squadron would specialize in the "Sandy" mission of rescuing downed pilots and aircrew. A-7D s/n 71-0309 carries LAU-10 rocket pods on station #1, a CBU-30 dispenser on #2, and Aero 1D fuel tank on #3. *Author's collection*

Third TFS aircraft sit in their revetments at Korat, circa 1973. Note the VA-147 A-7E in the background, left. *Author's collection*

The "Golden Dragons" of VA-192 would make three combat cruises to Yankee Station, all aboard *Kitty Hawk* as part of CVW-11. A 192 ordie cuts arming wires on Mk. 82s in preparation of another sortie, circa 1972–73. *Author's collection*

In May 1972, attack aircraft from CVW-15 took part in Operation Pocket Money, which called for the mining of Haiphong Harbor. VA-94's Hoboken (A-7E BuNo 157444) launches from *Coral Sea*, loaded with two Mk. 52 and two Mk. 36 mines. VA-94 and VA-22 would have the honor of being the last Navy A-7 squadrons to conduct combat sorties in support of the Vietnam War. *Author's collection*

In response to the Iranian hostage crisis, the US military planned a rescue effort known as Operation Eagle Claw. To ensure no confusion with Iraqi aircraft (then consisting of F-4s and F-14s), US naval aircraft had their starboard wings painted with a high-visibility stripe. Here an A-7E from VA-97 (BuNo 157528) launces from the USS *Coral Sea*, sometime in 1980. Sadly, Eagle Claw was aborted when lead element aircraft collided in the desert. *Naval Aviation Museum*

Toward the end of their tenure at Korat, A-7Ds of the 3rd TFS would wear black tailc odes instead of white. A-7D s/n 71-0315 prepares for a mission in July 1975. A few months earlier it had taken part in operations that provided air cover during the evacuations of Phnom Penh and Saigon, as well as the *Mayaguez* recovery. In December 1975, the squadron would leave Korat, bound for stationing at Clark AFB in the Philippines. *Author's collection*

In October 1983, USS *Independence* was called to Operation Urgent Fury, Grenada. A-7Es from VA-15 and VA-87 flew sorties in support of this operation. Warparty 401 (A-7E BuNo 156807) goes feet dry over Grenada's Saint-Georges Airport. *Photo by David Schneiderman*

In March 1986, A-7Es from CVW-17 onboard the *Saratoga* conducted strikes over Libya in retaliation of that country launching SA-5s at Navy F-14s during the Gulf of Sidra conflict. VA-81 aircraft served as decoys, flushing out Libyan SAM sites that were then hit by VA-83 aircraft carrying HARM missiles. It was the first time the HARM was fired in anger. Ram 310 (A-7E BuNo 160717) prepares for a HARM mission at Fallon in November 1986. *Naval Aviation Museum*

Having completed operations in Grenada, *Independence* and its CVW-6 airwing headed to the Med. In December, it would conduct combat operations in support of retaliatory strikes in Lebanon. Here, Echoes of VA-15 and VA-87 are spotted on the bow, most already loaded with Rockeyes. VA-15 would lose one aircraft to AAA fire during operations. The pilot was rescued. *Author's collection*

On December 20, 1989, A-7Ds from the Ohio ANG provided air support of the American invasion force tasked with capturing Panama's entrenched former president, Manuel Noriega. The 184th TFG had been on a planned deployment to Howard AFB as part of Operation Coronet Cove. *Author's collection*

CHAPTER 6

A-7 Operational Squadrons

US Air Force

Wing/Group	Squadron	ANG State	Tail Code(S)	Base City	Received	Relinquished	Notes
4525th Fighter Weapons Wing			WA	Nellis AFB, NV	Sep-69	Jun-72	Aircraft transferred to 66th FWS
57th Fighter Weapons Wing	66th Fighter Weapons Squadron		WA	Nellis AFB, NV	Jun-72	1981	Aircraft transferred to ANG
58th Fighter Training Wing	310th Tactical Fighter Squadron	Top Hats	LA	Luke AFB, AZ	Dec-69	Jul-71	Aircraft transferred to 333rd TFS
354th Tactical Fighter Wing	353rd Tactical Fighter Squadron		MR/MB	Myrtle Beach AFB, SC	Jul-71	1976	Aircraft transferred to ANG
354th Tactical Fighter Wing	355th Tactical Fighter Squadron	Fighting Falcons	MB	Myrtle Beach AFB, SC	Dec-70	1977	Aircraft transferred to ANG
354th Tactical Fighter Wing	356th Tactical Fighter Squadron		MN/MB	Myrtle Beach AFB, SC	May-71	1977	Aircraft transferred to ANG
354th Tactical Fighter Wing	511th Tactical Fighter Squadron		MR	Myrtle Beach AFB, SC	Jun-70	Jul-71	Aircraft transferred to 353rd TFS
354th Tactical Fighter Wing	4554th Tactical Fighter Replacement Squadron		MB	Myrtle Beach AFB, SC	May-72	1974–75	Aircraft transferred to ANG
355th Factical Fighter Wing	354th Tactical Fighter Squadron		DA/DM	Davis-Monthan AFB, AZ	Jul-71	1979	Aircraft transferred to ANG
355th Factical Fighter Wing	333rd Tactical Fighter Squadron		DM	Davis-Monthan AFB, AZ	Jul-71	1976	Aircraft transferred to ANG
355th Factical Fighter Wing	357th Tactical Fighter Squadron		DC/DM	Davis-Monthan AFB, AZ	May-71	1976	Aircraft transferred to ANG
355th Factical Fighter Wing	358th Tactical Fighter Squadron		DD/DM	Davis-Monthan AFB, AZ	Oct-71	1978	Aircraft transferred to ANG
23rd Tactical Fighter Wing	74th Tactical Fighter Squadron		EL	England AFB, LA	Jul-72	1981	Aircraft transferred to ANG
23rd Tactical Fighter Wing	75th Tactical Fighter Squadron		EL	England AFB, LA	Jul-72	1981	Aircraft transferred to ANG
23rd Tactical Fighter Wing	76th Tactical Fighter Squadron		EL	England AFB, LA	Oct-72	1981	Aircraft transferred to ANG
388th Tactical Fighter Wing	3rd Tactical Fighter Squadron		JH	Korat Royal Thai AFB, Thailand	Mar-73	1975	Aircraft transferred to ANG
4450th Tactical Group	4451st Tactical Squadron		LV	Nellis AFB, NV	1979	1989	8 A-7Ds and 2 A-7Ks as part of the F-117A program; aircraft transferred to ANG

A 76th TFS Big D (A-7D s/n 70-0955) heads toward the range, with Mk. 83 Snakeyes mounted to the parent wing stations. 23rd TFW aircraft often sported the shark's mouth reflective of their Flying Tigers' heritage. *Author's collection*

A three-ship of 354th Ds, all equipped with rocket pods, circa 1972–73. A-7D s/n 70-0970 now resides at the USAF Museum in Dayton, so honored as it was the Big D in which Maj. Colin Clarke flew a nine-hour Sandy on November 18, 1972, earning him the Air Force Cross, the service's second-highest honor. *Author's collection*

Wearing a short-lived, all-gray paint scheme, A-7D s/n 73-0993 sits on the ramp at Davis Monthan, May 11, 1974. Upon retirement of the type from the 355th, it would go to the Pensylvania ANG and serve with the 112th TFS until eventual retirement. *Photo by Ben Knowles, David F. Brown collection*

A couple of 3rd TFS Big Ds on the ramp at Korat, date unknown. A-7D s/n 71-0315 sits in the foreground, loaded with LAU-10 rocket pods on station #3, a CBU-30 on #2, and Aero 1D tank on #3. *Author's collection*

In addition to conducting tests of new equipment and modifications, the 6510th also operated a few A-7Ds for the base's Test Pilot School. TPS aircraft were painted in high-vis white and red markings, as seen here on A-7D s/n 69-6194 as it flies formation on 69-6217, August 8, 1984. *Photo by Keith Svendsen*

Upon delivery to Nellis, the 4450th applied Euro 1 paint schemes to their A-7s, as well as the LV tailc ode. Official story at the time was that the squadron was undertaking "avionics testing." In fact, the squadron's aircraft were providing the group with a plausible cover story while they secretly operated F-117A stealth fighters. *Author's collection*

Air National Guard

Wing/Group	Squadron	ANG State	Tail Code(S)	Base City	Received	Relinquished	Notes
112th Tactical Fighter Group	146th Tactical Fighter Squadron	PA	PT	Pittsbugh	1980	1991	Received A-7K 1983
114th Tactical Fighter Group	175th Tactical Fighter Squadron	SD	SD	Sioux Falls	1977	1992	Received A-7K 1982
121 Tactical Fighter Wing	112th Tactical Fighter Squadron	OH	OH	Toledo	1979	1992	Received A-7K 1980
121 Tactical Fighter Wing	162nd Tactical Fighter Squadron	OH	OH	Springfield	1978	1992	Received A-7K 1982
121 Tactical Fighter Wing	166th Tactical Fighter Squadron	OH	OH	Columbus	1974	1992	Received A-7K 1983
127th Tactical Fighter Wing	107th Tactical Fighter Squadron	MI	MI	Detroit	1977	1992	Received A-7K 1981
132nd Tactical Fighter Wing	124th Tactical Fighter Squadron	IA	IA	Des Moines	1977	1992	Received A-7K 1981
138th Tactical Fighter Group	125th Tactical Fighter Squadron	OK	TK/OK	Tulsa	1981	1993	Received A-7K 1983
140th Tactical Fighter Wing	120th Tactical Fighter Squadron	CO	CO	Aurora	1974	1991	Received A-7K 1983
150th Tactical Fighter Wing	188th Tactical Fighter Squadron	NM	NM	Albuquerque	1974	1992	
156th Tactical Fighter Wing	198th Tactical Fighter Squadron	Puerto Rico	PR	San Juan	1975	1991	Received A-7K 1983
162nd Tactical Fighter Group	152nd Tactical Fighter Squadron	AZ	AZ	Tucson	1978	1986	Received A-7K 1983
162nd Tactical Fighter Group	195th Tactical Fighter Squadron	AZ	AZ	Tucson	1983	1992	Received A-7K 1983
169th Tactical Fighter Group	158th Tactical Fighter Squadron	SC	SC	Columbia	1975	1983	Received A-7K 1982
185th Tactical Fighter Group	174th Tactical Fighter Squadron	IA	HA	Sioux City	1976	1991	Received A-7K 1983
192nd Tactical Fighter Group	149th Tactical Fighter Squadron	VA	VA	Richmond	1981	1992	Received A-7K 1982
4450th Tactical Group	4451st Tactical Squadron		LV	Nellis AFB, NV	1979	1989	8 A-7Ds and 2 A-7Ks as part of the F-117A program; aircraft transferred to ANG

Showing beautifully against the York County, Pennsylvania, countryside, A-7D s/n 74-0742 carries an Aero 1D tank on station #3, and empty TER on #2. Photo taken in November 1989. The squadron retired their A-7s in 1991, and with this came a change of mission, the 146th now flying KC-135s in the refueling role. *Photo by David F. Brown*

An A-7K (s/n 80-0292) assigned to the 175th Tactical Fighter Squadron, 114th Tactical Fighter Group of the South Dakota Air National Guard. The pod on the outboard station is a blivet; basically a luggage container used on cross-country flights. This aircraft was lost on May 30, 1990, when it collided with an A-7D over Spencer, Iowa. *Author's collection*

Wearing a Euro 1 scheme, a 112th TFS A-7D (s/n 70-0894) delivers a parachute-retarded Mk. 82, circa 1989. *Author's collection*

The 120th TFS received their A-7K in 1981. A-7K s/n 81-0077 departs Des Moines International circa 1987, carrying a FLIR pod on station #6. 0077 was the very last A-7K built and served exclusively with the 13nd TFW until its retirement in 1992. *Author's collection*

A-7D 70-0991 taxis at Harrisburg International Airport, Pennsylvania, September 4, 1987. With the introduction of the Euro 1 paint scheme, the 127th painted both the ANG crest (on tail) and Wing crest (forward of the wing) in dark-red outlines. Note the personalized blivet on station #1. *Photo by David F. Brown*

It was quite common for ANG squadrons to deploy overseas on exercises. Here A-7D s/n 69-6218 sits on the ramp at Volkel Air Base in the Netherlands on April 23, 1985. By this time the squadron dropped the "TL" tail code for "OK." *Photo by Rick Versteeg*

Toward the end of their careers in the ANG, most A-7s would be painted in a two-tone-gray, wraparound scheme like the one featured here on s/n 69-6242. The CO flag motif on the tail has been replaced by a stylized Rocky Mountain horizon, with scripted "Colorado." It was also given the nickname "Rocky Mountain Thunder," seen to the right of the squadron insignia. *Author's collection*

A-7D s/n 72-0237 shows off revised tail art of a yellow Zia (the state's symbol) with a roadrunner perched inside of it. Photographed on its home ramp in Albuquerque in March 1981. *Bob Lawson Collection, Naval Aviation Museum*

Back on its home ramp at Muñiz ANG Base, A-7D s/n 69-6226 receives some TLC from airmen. The waist-high accessibility to a lot of the A-7's innards made for appreciative maintainers. The intake is covered with a high-speed run-up FOD protector. *Author's collection*

In the fall of 1981 the 162nd helped the Air Force experiment with a number of different paint schemes, testing them against a variety of terrains to determine effectiveness. *Author's collection*

The 169th TFW was the first Air National Guard unit to retire their A-7s, doing so in 1983. Here s/n 73-1010 sports a wraparound SEA scheme, date and location unknown. It would go on to serve the rest of its career with the Virginia ANG and is currently on display at a veteran's memorial in Alexander City, Alabama, erroneously marked in Alabama ANG colors. *Author's collection*

In 1977, the "Bats" of the 174th TFS traded in their F-100s for A-7Ds. Squadron aircraft are seen here on a deployment to Nellis in April 1981, with s/n 69-1235 in the foreground. Note the yellow squadron tail stripe, which includes a bat's silhouette. The following year this aircraft would be reassigned to the Puerto Rico ANG. *Photo by Keith Svendsen*

A-7D s/n 70-0982 banks away from the photo ship, showing off a TER of Mk. 76 practice bombs. In 1991, this aircraft was transferred to the Puerto Rico ANG. While on a training deployment to Volk Field, Wisconsin, cracks in the wing were discovered—something not too uncommon to A-7Ds toward the end of their service life. 0982 stayed at Volk, where it serves as a gate guardian. *Photo by Barry Roop*

US Navy

Squadron	Call Sign(s)	Home Port	Variant	Received	Retired
VA-12 Flying Ubanis (up to 1982); Clinchers (1982 on)	Skull (at sea); Clincher (shore)	NAS Cecil Field, FL	A-7E	Apr-71	Oct-86
VA-15 Valions	Active Boy	NAS Cecil Field, FL	A-7B	Mar-69	Aug-75
			A-7E	Aug-75	Dec-86
VA-22 Fighting Redcocks	Beefeater	NAS Lemoore, CA	A-7E	Feb-71	Jul-90
VA-25 Fist of the Fleet	Canasta	NAS Lemoore, CA	A-7B	Oct-68	Dec-69
			A-7E	Dec-69	May-83
VA-27 Royal Maces	Mace/Charger	NAS Lemoore, CA	A-7A	Jan-68	Jun-70
			A-7E	Jun-70	Dec-90
VA-37 Bulls	Falcon	NAS Cecil Field, FL	A-7A	Aug-67	Mar-73
			A-7E	Apr-73	Oct-90
VA-46 Clansmen	Chief	NAS Cecil Field, FL	A-7B	Nov-68	Jul-77
			A-7E	Jul-77	Jun-91
VA-56 The Champions	Champ	NAS Yokosuka (73); NAS Lemoore (86)	A-7B	Jan-69	Mar-73
			A-7A	Mar-73	Apr-77
			A-7E	Apr-77	Aug-86
VA-66 Waldoes	Waldo	NAS Cecil Field, FL	A-7E	Oct-70	Oct-86
VA-72 Blue Hawks	Decoy	NAS Cecil Field, FL	A-7B	Jan-70	Sep-77
			A-7E	Sep-77	Jun-91
VA-81 Sunliner	Sunliner	NAS Cecil Field, FL	A-7E	May-70	Feb-88
VA-82 Marauders	Streetcar	NAS Cecil Field, FL	A-7A	Jun-67	Sep-70
			A-7E	Sep-70	Apr-72
			A-7C	Apr-72	Nov-74
			A-7E	Nov-74	Nov-87
VA-83 Rampagers	Ram	NAS Cecil Field, FL	A-7E	Jun-70	Mar-88
VA-86 Sidewinders	Winder	NAS Cecil Field, FL	A-7A	Feb-67	Oct-70
			A-7E	Oct-70	Apr-72
			A-7C	Apr-01	Nov-74
			A-7E	Nov-74	Jul-87
VA-87 Golden Warriors	Warparty	NAS Cecil Field, FL	A-7B	Jun-68	Oct-75
			A-7E	Oct-75	Oct-86
VA-93 Blue Blazers (before 76); Ravens (76 on)	Raven	NAS Yokosuka (73); NAS Lemoore (86)	A-7B	Apr-69	Mar-73
			A-7A	Mar-73	Apr-77
			A-7E	Apr-77	Aug-86
VA-94 Mighty Shrikes	Hoboken	NAS Lemoore, CA	A-7E	Jan-71	Apr-90
VA-97 Warhawks	Warace	NAS Lemoore, CA	A-7A	Oct-67	Jul-70
			A-7E	Jul-70	Dec-90

Squadron	Call Sign(s)	Home Port	Variant	Received	Retired
VA-105 Gunslingers	Canyon Passage	NAS Cecil Field, FL	A-7A	Nov-67	Feb-73
			A-7E	Sep-74	Apr-90
VA-113 Stingers	Stinger	NAS Lemoore, CA	A-7B	Dec-68	Apr-70
			A-7E	Apr-70	Jul-83
VA-122 Flying Eagles	Eagle	NAS Lemoore, CA	A-7A	Nov-66	
			A-7B	May-68	
			A-7E	Jul-69	May-91
			A-7C	Jul-71	
			TA-7C	1978	May-91
VA-125 Rough Raiders	Raider	NAS Lemoore, CA	A-7B	Sep-69	Oct-77
			A-7A	Oct-69	Sep-69
			A-7C	Aug-75	Oct-77
VA-146 Blue Diamonds	Diamond	NAS Lemoore, CA	A-7B	Jun-68	Sep-69
			A-7E	Sep-69	Jul-89
VA-147 Argonauts	Jason	NAS Lemoore, CA	A-7A	Jun-67	Sep-69
			A-7E	Sep-69	Jul-89
VA-153 Blue Tail Flies	Saddleback	NAS Lemoore, CA	A-7A	Sep-69	May-73
			A-7B	May-73	Sep-77
VA-155 Silver Foxes	Powerhouse	NAS Lemoore, CA	A-7B	10.69	Sep-77
VA-174 Hell Razors	Razor	NAS Cecil Field, FL	A-7A	Oct-66	
			A-7B	Jul-68	
			A-7E	Dec-69	Jun-88
			TA-7C	Jun-78	Jun-88
VA-192 Golden Dragons	Dragon	NAS Lemoore, CA	A-7E	Feb-70	Oct-86
VA-195 Dambusters	Chippie	NAS Lemoore, CA	A-7E	Feb-70	Apr-85
VA-215 Barn Owls	Barn Owl	NAS Lemoore, CA	A-7B	Oct-69	Sep-77
VAQ-33 Firebirds	Firebird	NAS Key West, FL	TA-7C		Oct-93
			EA-7L		Oct-93
VAQ-34 Flashbacks	Flashback	NAS Pt. Mugu, CA; NAS Lemoore (6/91)	TA-7C		Oct-93
			EA-7L		Oct-93
VX-5 Vampires	Vampire	NAS China Lake, CA	A-7A	Dec-66	
			A-7B	Apr-68	
			A-7E	Apr-69	
			A-7C	Sep-71	
			TA-7C	Jun-78	

VA-12 was reestablished as an A-7E squadron in April 1971 and flew the variant for the next fifteen years. The squadron made their first A-7 cruise in September 1971 aboard the *Independence* with its fellow CAG-7 squadrons. For that cruise, the "Ubangis" wore the 500 modex. Here's Skull 500 (A-7E BuNo 157563), date and location unknown. Note the protective, anti-FOD intake screen, used on ground engine run-ups. *Bob Lawson Collection, Naval Aviation Museum*

Wearing the original squadron markings, Pride 303 (A-7B BuNo 154412) flies wing on Pride 301 (A-7B BuNo 154415), circa 1970. *Naval Aviation Museum*

VA-22 markings were fairly consistent for most of the squadron's tenure with the A-7E. Here Beefeater 302 (A-7E BuNo 159273) provides a nice view of the A-7's planform. Photo taken over the California countryside, sometime in 1979. *Author's collection*

The "Fisties"' CAG bird launches from *Ranger*, sometime in 1981. The squadron started toning down their tail colors, removing the all green but still retaining the squadron emblem. The insignia on the bottom of the rudder is that of CVW-2. Note that the UHTs are fully deflected, and the spoiler on the top of the starboard wing is already engaged, indicating a climbing, clearing turn to the left upon launch. *Author's collection*

Having just trapped, NK-403 (A-7E BuNo 156808) is directed out of the landing area on the USS *Enterprise*, sometime in 1975. This Echo would finish up its career serving in VA-205, the Atlanta-based Reserve squadron. *Naval Aviation Museum*

Beautiful lighting captures Falcon 313 as it sits on the deck of the USS *Saratoga,* sometime in the mid-1970s. The aircraft is loaded out with a single Mk. 82 on station #1, and an LB-31A camera pod on station #3. *Author's collection*

Starting life in A-4s, the "Clansmen" of VA-46 got their A-7Bs in November 1968 and flew the variant for nearly eleven years. Here, Chief 306 (A-7B BuNo 154487) sits on the ramp at its homep ort, NAS Cecil Field. *Naval Aviation Museum*

VA-56's Champ 407 (A-7B 154362) at NAS Lemoore, sporting the markings of CVW-16. Soon thereafter, their modex would go from "AH" to "NF", being reassigned to CVW-5. Photo taken January 1971, photographer unknown. *Naval Aviation Museum*

Champ 404 (A-7E BuNo 156858) sporting interim, low-visibility markings. VA-56 was one of the few squadrons who felt the need to paint the exhaust of the environmental control system red (seen below the US insignia). This photo was taken sometime in 1980; the photographer is unknown. *Naval Aviation Museum*

A three-ship of "Waldos" over the Atlantic, circa 1980. By this time, the squadron incorporated a yellow, circular addition to their tail stripe that featured the CAG-7 "AG" modex. *Author's collection*

A breathtaking shot of two VA-72 Echoes taken near Gibraltar, circa 1980. *Author's collection*

In the early 1980s, the "Sunliners" were sporting a new scheme, as well as a new piece of equipment—the FLIR pod. The first East Coast squadron so equipped, Sunliner 404 (A-7E BuNo 160718) shows off its lines sometime in 1981. Note that the squadron's large, orange zapper has been reduced, showing up on the fin cap. *Author's collection*

Streetcar 301 (A-7E BuNo 159301) photographed October 1977 on board the USS *Nimitz*. Markings at the time retained the stylized eagle, with a new lightning bolt. This aircraft is on static display at the Oakland Aviation Museum, wearing USAF A-7D colors. *Naval Aviation Museum*

A few years later, VA-83 markings changed again, trading in the ram's head for a charging ram that was appropriated from a Dodge promotion. Here, Ram 304 (A-7E BuNo 160867) traps aboard *Forrestal,* April 1982. This aircraft would later serve with VA-72 during Operation Desert Storm and would end up its service life with the Hellenic air force. *Photo by Perry Coley, Naval Aviation Museum archives*

Seen here on a mission during the squadron's last combat cruise, Winder 413 (A-7C BuNo 156760) shows off new markings, which included a diamond on the rudder, as well as a large, dorsal diamond immediately behind the cockpit. The squadron insignia has returned, this time on the leading edge of the tail. Photo was taken sometime in 1973. *Author's collection*

By 1973, VA-87 aircraft were sporting red instead of blue and added a colorful warbonnet to its tail art. Also of note is the use of a tomahawk (immediately above the intake warning triangle), on which squadron pilots' names would be featured. Here, Warparty 404 (A-7B BuNo 154440) flies wing on 401 (A-7B BuNo 154471) over a European coastline, sometime in late 1973 or early 1974. *Naval Aviation Museum*

Now sporting the larger shark's mouth that became the squadron's standard, Raven 315 (A-7A BuNo 153173) is readied on the deck of the USS *Midway*. It's configured for a mining exercise in the South China Sea, conducted on March 24, 1976. *Photo by Bob Thomas*

Captured in 1977–78, VA-94's Hoboken 410 (A-7E BuNo 156828) flies wing on 401 (A-7E BuNo 159981) somewhere over the Pacific. *Author's collection*

By the late 1980s, the *Mighty Shrikes* were wearing subdued, tactical gray markings. Hoboken 407 (A-7E BuNo 160731) is seen on the ranges at Fallon, dropping Mk. 82 Snakeyes. This aircraft would later be assigned to VA-122, where it was lost in a midair collision in October 1987. The pilot was able to eject safely. *Naval Aviation Museum*

By their fourth Vietnam cruise, the "Warhawks" had traded in their Alphas for Echoes and painted their radomes black. Here Warace 301 (A-7E BuNo 156846) heads north on a mission. *National Naval Aviation Museum*

VA-105's CAG bird (A-7E BuNo 159970) seen at Fallon in November 1983, shortly after returning from its world cruise on *Carl Vinson*; hence the "NL" modex. It carries an inert, training AGM-45 Shrike missile on station #8. *Photo by Michael Grove, Naval Aviation Museum archives*

Photographed in her final paint scheme, Canyon Passage 410 (A-7E BuNo 157569) shows off her lines, circa 1989. The squadron traded in their Echoes for Hornets in April 1990. *Author's collection*

This unique view shows just how roomy the A-7 cockpit was. Photo of a VA-122 Echo conducting carrier qualifications aboard the USS *Ranger*, in September 1987. *Photo by Rick Morgan*

VA-122 received their first TA-7Cs in 1978. Here, Eagle 206 (TA-7C BuNo 154412) flies lead with 201 (A-7E BuNo 156841), with a load of "blue death," a.k.a. Mk. 76 practice bombs, circa 1979. *Author's collection*

Stinger 304 (A-7E BuNo 158661) heads out from Fallon with a load of inert Mk. 82s. Interesting to note are the three red warning arrows surrounding the environmental control unit exhaust—markings unique to the squadron. *Naval Aviation Museum*

A four-ship of Rough Raider Bravos over Fresno, California, February 15, 1970. VA-125 assumed the role of West Coast RAG for the A-7A/B community, allowing VA-122 to focus on the A-7E. *Naval Aviation Museum*

Just having trapped aboard the USS *Constellation,* Diamond 303 (A-7E BuNo 158278) begins to fold its wings as it taxis out of the landing area, February 24, 1978. By the mid-1970s, squadron aircraft were sporting this eye-catching scheme. This aircraft is on display at the Southern Museum of Flight in Birmingham, Alabama. *Photo by Pete Clayton, Naval Aviation Museum*

Taken on the ramp at Fallon in January 1975, VA-147's CAG bird (A-7E BuNo 156834), a.k.a. "City of Sacramento," sports squadron markings all around—on the tip of the TER rack and LAU-7 Sidewinder rail, as well as the Aero 1D fuel tank. *Photo by Michael Grove, Naval Aviation Museum*

Onb oard for its final cruise on *Oriskany,* Saddleback 307 (A-7B 154400) is decked out in bicentennial tail markings. This last cruise on *Oriskany* concluded on March 3, 1976, so it is assumed this photo was taken prior to that time. *Photo by Pete Clayton, Naval Aviation Museum*

Another bicentennial bird, this one Powerhouse 501 (A-7B BuNo 154456), rides on its nose gear as it traps on *Oriskany*, sometime in 1976. Like its cousin, the F-8 Crusader, the A-7 tended to "dance" on its nose gear when landing, due to the forward momentum, sudden deceleration, and a forward center of gravity. Recognizing this, Vought designed the A-7 with dual nose gear tires, providing more stability as well as less tendency for tire blowouts—something the F-8 was (in)famous for. *Author's collection*

A four-ship of very early Alphas, circa 1966–67. Note that all aircraft feature full BuNos on their tails, and the "Hell Razor" insignia on the fuselage, immediately behind the yellow rescue arrow. The insignia would soon go away, and the tail BuNos would be shortened to the last four numbers, but little else would change over the squadron's tenure. *Author's collection*

In 1976, the squadron replaced the dragon on the forward fuselage (detractors would refer to it as "the worm") in favor of a circular design similar to the squadron's insignia. Here, Dragon 303 (A-7E BuNo 157524) taxis out at Fallon in May 1977 with a TER load of practice Snakeyes. *Naval Aviation Museum*

The original Chippie paint scheme featured a larger eagle's head than would be seen in later versions. Photographed in front of the hangar at Lemoore shortly after assignment in February 1970, Chippie 401 (A-7E 156880) would be reassigned to VA-94 before the "Dambuster's inaugural A-7 cruise in November of that year. It would be lost in January 1972, when the 94 pilot was forced to eject after losing his starboard main gear on launch from *Coral Sea. Naval Aviation Museum*

VA-215 would be disestablished in September 1977, but not before it made its one and only cruise on the *FDR*. Barn Owl 404 (A-7B BuNo 154411) is captured here on the Lemoore ramp in June 1977. It would finish its service in the Reserves, flying with VA-305 out of NAS Point Mugu, California. *Naval Aviation Museum*

Firebird 114 (TA-7C BuNo 156747) at home in Key West, carrying a DLQ-3 pod. The pod was used to generate noise and deception jamming, used for radar and missile system tests and evaluations. When VQ-33 was disestablished in October 1993, this Twosair was renovated in Jacksonville and sold to the Greeks. *Author's collection*

A nice VAQ-34 two-ship, showing the squadron's original high-vis and transitional low-vis schemes. Both aircraft carry an AST-4 threat simulator pod on station #8, and Aero 1D on #6. *Author's collection*

The Naval Air Test Center operated a number of A-7s during the type's use in service, most sporting high-vis red noses, tails, and wingtips. Here, the second A-7A prototype, BuNo 152581, is seen on static display during a Pax River airs how, September 20, 1975. *Naval Aviation Museum*

A-7s were assigned to the Naval Weapons Evaluation Facility in Albuquerque, New Mexico. These aircraft were used primarily as carriers for evaluating changes made to weapons and missiles, ensuring safe release and accurate delivery. A-7C BuNo 156772 was photographed at NAS Miramar, June 8, 1978. Note the camera mounting near the intake. *Photo by Ray Lock, Naval Aviation Museum*

The Pacific Missile Test Center (PMTC) at Pt. Mugu also utilized A-7s in the role of launch platforms. A-7A BuNo 152651 taxis in after a test, May 20, 1977. Note the camera mount on the fuselage, just above the nose gear door. *Photo by Bob Lawson*

The Naval Weapons Center (NWC) at China Lake flew their share of A-7s. Here two Echoes, BuNo 160566 (foreground) and BuNo 156883 (behind) drop Mk. 82 Snakeyes on the range, circa 1980. Each carries an AN/AAR-45 FLIR (Forward-Looking Infrared) pod on station #6. NWC helped to develop the FLIR pod, as well as tactics for its use. The tail art of both Echoes feature the "Night Raider" insignia developed for this project. *Author's collection*

US Naval Reserve

Squadron	Name	Call Sign	Home Port	Variant	Received	Retired
VA-203	Blue Dolphins	DOLPIN	NAS Cecil Field, FL	A-7A	Apr-74	Aug-77
				A-7B	Aug-77	Sep-83
				A-7E	Sep-83	Oct-89
VA-204	River Rattlers	River	NAS New Orleans, LA	A-7B	Mar-78	Jun-86
				A-7E	Jun-86	Dec-90
VA-205	Green Falcons	Salty	NAS Atlanta, GA	A-7B	Sep-75	Jun-84
				A-7E	Jun-84	Aug-90
VA-303	Golden Hawks	Hawk	NAS Alameda, CA	A-7A	Apr-74	Aug-77
				A-7B	Aug-77	Oct-85
VA-304	Firebirds	Firebird	NAS Alameda, CA	A-7A	Aug-71	Sep-77
				A-7B	Sep-77	Sep-86
				A-7E	Sep-86	Aug-88
VA-305	Lobos	Lobo	NAS Pt. Mugu, CA	A-7A	Apr-71	Aug-77
				A-7B	Aug-77	Dec-86

Hellenic Air Force (Greece)

Squadron	Name	Home Base	Variant	Received	Retired
335th Squadron	Tigers	Araxos	A-7H/TA-7H	1992	1993
			A-7E/TA-7C	1993	2007
336th Squadron	Olympos	Araxos	A-7E/TA-7C	1993	2014
			A-7H/TA-7H	2002	2007
340th Squadron	Foxes	Souda	A-7H	1975	2001
			TA-7H	1980	2001
345th Squadron	Laelaps	Souda	A-7H	1976	2002
			TA-7II	1980	2002
347th Squadron	Perseus	Larissa	A-7H	1977	1992
			TA-7H	1980	1992

Portuguese Air Force

Squadron	Name	Home Base	Variant	Received	Retired
302 Squadron	Falcoes	Air Base No. 5, Monte Real	A-7P	1981	1996
			TA-7P	1985	1996
304 Squadron	Magnificos	Air Base No. 5, Monte Real	A-7P	1984	1999
			TA-7P	1985	1999

Royal Thai Navy

Squadron	Name	Home Base	Variant	Received	Retired
104th Squadron	White Sharks	U-Tapao Air Base	A-7E	1995	2007
			TA-7C	1995	2007

CHAPTER 7

A-7s for Export

Knowing they had a winner in the A-7, Vought was keen to solicit foreign sales. The US State Department, on the other hand, didn't necessarily share Vought's enthusiasm. Given its long range and state-of-the-art weapon systems, the State Department understood that the A-7 would be a game changer, potentially shifting the balance of power in any region in the favor of the country that deployed them. As a result, export numbers for the A-7 were perhaps not as strong as they potentially could have been. Vought did have some success in selling the A-7 to other countries, and those who operated them loved them.

Greece

The Hellenic air force (HAF) was by far the most important operator of the A-7 outside the US, purchasing nearly 10 percent of the total number of A-7s produced. In June 1974, the Greeks signed a contract for sixty aircraft, designated the A-7H. The H model was essentially an A-7E, with the addition of the internal jet fuel starter, like that used on the A-7D. The more obvious differences in the A-7H and Echo model was the removal of the launch bar and refueling probe. The later omission was mandated by the US government to placate Turkish concerns that their neighbors (and frequent foes) were getting the A-7.

In January 1975, HAF pilots and personnel arrived at NAS Cecil Field for familiarization and training with the "Hell Razors" of VA-174, the East Coast A-7 RAG. Delivery of the first A-7H—BuNo 159662—was made in May of that year and continued apace over the next five years. By July 1975, having returned from Cecil, the newly trained Greeks reported to 340 Squadron at Souda Air Base and set about trading in the unit's venerable F-84Fs for the arriving A-7Hs. In July 1976, 345th Squadron were also outfitted with A-7Hs, sharing the ramp at Souda with the 340th.

It was also in 1976 that HAF pilots undergoing training at Cecil Field had the opportunity to check out the new TA-7C. Recognizing the value of a two-seater, the Greeks requested a purchase of five to be built to A-7H standards. These aircraft were delivered over the course of 1980 and were assigned to the 115 Combat Wing, who provided the necessary training for 340 and 345 Squadrons. Prior to receiving their Twosairs, the 115CW used T-33s in the training of A-7H pilots.

The A-7H was well suited for the HAF, and it would serve in a total of five squadrons before their retirement in 2007, by then having flown in excess of 320,000 hours. Faced with a need to find a replacement for their aging F-104s, and wishing to bolster their fleet of A-7Hs, the Greeks contracted for the acquisition of retired US Navy A-7Es and TA-7Cs. Some of these aircraft would come from the ranks of Desert Storm veterans of VA-46 and VA-72, which had recently been deactivated. Others would be selected from A-7s in storage at AMARG in Tucson. A total of sixty-eight airframes would be purchased, with thirty-two Echoes and four Twosairs being refurbished in Jacksonville and flown over to Araxos Air Base. The first of these ferry flights arrived to Araxos on March 25, 1993. The balance of the aircraft was shipped over to Greece, where they were rebuilt and restored to flight on-site.

Like the A-7H/TA-7Hs before them, the A-7E/TA-7Cs served the HAF well. But, all good things must come to an end, and the Greeks retired their A-7 fleet in October 2014, after thirty-nine years of faithful service.

Portugal

The Portuguese air force (PoAF) had been in the market for a replacement to their F-86F Sabres since 1968. While initial considerations called for the acquisition of both a fighter/interceptor and a ground attack/support aircraft, economic realities constrained Portugal's search to just one aircraft. In 1974, Portugal reached an agreement with the US, providing them use of Lajes Air Base in the Azores. One advantage to this agreement was that PoAF's purchase of aircraft could be offset by the lease of Lajes to the Americans—a lease estimated to be worth $72 million. Having looked at a number of options, the PoAF began to see the A-7 as a sound solution. The income that would come from Lajes would not only pay for A-7s aircraft but cover the cost of modernizing and supporting them as well.

On May 5, 1980, a contract was inked calling for twenty aircraft, now designated A-7P. These aircraft had begun their lives as A-7As, with a few being veterans of VA-147's first Vietnam combat cruise. The airframes were transported by rail to the Vought factory, where they underwent extensive restoration and upgrading. The A-7P would retain the TF-30-P-408 engine, as well as the Alpha's two Mk. 12 cannons. The avionics suite featured improvements common to both A-7E and D variants.

Official acceptance of the A-7P was marked with a ceremony at Andrews Air Force Base on August 19, 1981, where the Portuguese ambassador confirmed his country's plans to acquire more aircraft. The initial cadre of PoAF pilots and maintainers arrived at the Vought facility in October 1981 to undergo training. In late December, the first A-7Ps would transit from the Vought factory to Monte Real Air Base, where they were assigned to the Falcöes of 304 Squadron. In May 1983, an order for thirty aircraft was finalized, calling for twenty-four A-7Ps and six two-seat TA-7Ps. This contract necessitated the need for a second operator at Monte Real, the Magnificos of 302 Squadron.

While the A-7P represented a huge leap forward in technology and capability for the PoAF, their potential was plagued by ongoing budget constraints, lack of spare parts, and a high accident rate. All told, sixteen aircraft were lost—a third of the fleet—resulting in the death of six pilots and one maintainer. By the late 1980s, the PoAF had set its sights on acquiring the F-16. 302 Squadron was deactivated, with its pilots and personnel being merged into 304 Squadron, which itself was deactivated on July 9, 1999.

Thailand

In 1993, the Royal Thai Navy (RTN) decided to outfit one squadron with A-7s for use in the coastal defense role and purchased fourteen ex-Navy A-7Es and six TA-7Cs for this purpose. The aircraft were selected from those in long-term storage in Tucson and then transported to the Naval Air Rework Facility (NARF) in Jacksonville, Florida, for restoration. After being completely overhauled and test-flown, the aircraft were ferried from Jacksonville to the RTN base at U-Tapao, with flights taking place in 1995. It's interesting to note that none of these ferry flights utilized in-flight refueling but instead relied on internal fuel as well as those carried in Aero 1D tanks. This resulted in six-day transits, each requiring nine stops en route, and roughly twenty-six flight hours.

The A-7s were operated by the 104th Squadron—the White Sharks. 104th pilots were provided necessary training by contracted instructors (most of them Navy A-7 veterans) both in the US and back in Thailand. Members of the 104th enjoyed flying and maintaining the A-7s and did so for roughly twelve years. Sadly, by 2007, severe budget constraints led to their being kept in nonflying-but-ready status. The aircraft were occasionally run up, and some even taxied, but they never took to the skies after 2007. All airframes were eventually disposed of, either to museums or on outdoor, static display.

Failed Export Efforts of Note

SWITZERLAND

In 1971, the Swiss air force, recognizing the need to replace their aging de Havilland Venoms, solicited proposals for a new aircraft. In the end, the finalists were the Fiat G.91, Douglas A-4M, Dassault's Mirage, and the Vought A-7. Vought's proposal—designated the A-7G—was based on the A-7D. Proposed changes included an uprated TF-41 and replacing the M61 Vulcan with two Swiss-built 23 mm cannons, as well as substituting the Loran and Doppler systems for Swiss equivalents. Two A-7Ds were sent to Switzerland as A-7G demonstrators, operating extensively throughout 1972 in a number of tests and fly-off competitions. Despite its strong showing, the Swiss not only passed on the A-7G, but all other finalists as well. In the end, they decided to purchase thirty refurbished Hawker Hunters as an interim solution, eventually acquiring the Northrop F-5 Freedom Fighter.

PAKISTAN

Having flown their F-86Fs for nearly twenty years, the Pakistan air force was similarly looking to upgrade. By 1974, the Pakistanis had narrowed their selections to either the A-7 or the F-5, with a clear preference for the Corsair II. In that same year a contingency of Pakistani officials visited the Vought factory in Texas to inspect and evaluate the A-7. Negotiations continued over the next three years, and Pakistan expressed its desire to purchase as many as 110 aircraft. The Carter administration feared that Pakistani A-7s would prove too much of a destabilizing factor in the region. Carter's diplomats did promise to approve the sale of A-7s, but only on the condition that Pakistan would scale back its nuclear efforts, rationalizing that the A-7 provided an overwhelming deterrent to neighboring India. In the end, Pakistan refused the offer and decided to acquire Chinese-built MiG-19s and, eventually, F-16s.

The 1974 contract called for sixty A-7Hs to be built, and production started later in the year. The final A-7H would leave the Vought factory in 1980. *Author's collection*

Nikolaos Papadopoulos prepares for his first flight in an A-7. Papadopoulos served as the first commander of 340 Squadron, the first to receive the A-7H. He would later become the first commander of 347 Squadron. *Author's collection*

Nikolaos Siganos (*right*), the commander of 345 Squadron, receives some instruction from Lt. Bob Lakari of VA-174. The "Hell Razors" were responsible for the initial training of all HAF pilots and maintenance personnel. *Author's collection*

The first A-7H (BuNo 159662) heads from the paint shed. Its main gear door carries a stencil that reads, "Fuel with JP5 Fuel Only." *Author's collection*

Cmdr. Papadopoulos discussion of the finer points of A-7 handling by an unidentified fellow HAF pilot. The lessons learned in the classroom and over the skies of Florida, would soon be shared with Papadopoulos' pilots in Greece. *Author's collection*

BuNo 159662 takes to the skies for its first flight, May 6, 1975. Vought's chief test pilot, John Konrad, again did the honors. It would be lost in an accident on September 3, 1977. The pilot was able to successfully eject. *Author's collection*

With the arrival of the A-7Es, the HAF added a very valuable skill to their A-7 repertoire, aerial refueling! Training was conducted from Araxos, with former Navy A-7 pilots providing the instruction in both day and night plugging. *Photo courtesy of Papilos Papadopoulos*

The A-7H was well loved by the HAF, and certainly well used, as evidenced by this photo of A-7H BuNo 159953. While similar to the A-7E, H models lacked the refueling probe, and in its place had a longitudinal "plug." The Greeks would operate the A-7H for thirty-two years. Photographed in May 1992, photographer unknown. *David F. Brown collection*

TA-7C BuNo 156774 flies lead on A-7E BuNo 160560 as they depart Araxos, May 4, 2014. The Twosair was mistakenly marked as BuNo 155774, and the error was quickly fixed. *Photo courtesy of Papilos Papadopoulos*

All of the A-7s destined for service in the PoAF began their life as A-7As. The airframes were transported from the Tucson desert to Vought, where they underwent extensive work to bring them up to A-7P standards. This A-7A (BuNo 153184) would serve as A-7P 5504 and remains on display at the Ovar Figo Maduro Air Base. *Author's collection*

In 2005, the HAF introduced the Archangel International Air Show, which they held at Tangara Air Base. For this event, 336 Squadron painted A-7E BuNo 158825 in a dramatic, Tiger paint scheme. The response was so positive that the aircraft was invited to participate in many air shows around Europe. In 2007, the squadron painted another Echo, this one BuNo 160616, in a black- and-silver "Olympos" scheme. Like Tiger, it took the air show community by storm. Here the two aircraft are seen flying together on January 28, 2006. *Photo courtesy of Papilos Papadopoulos*

The A-7P acceptance ceremony was held at Andrews AFB on August 19, 1981. A-7P 5501 (originally BuNo 154352) was on hand to provide dignitaries and guests the opportunity to inspect the newly acquired aircraft. 5501 would be lost over Belgium in a collision with A-7P 5505 (BuNo 153190) on July 1, 1985. *Author's collection*

On December 21, 1981, the first nine A-7Ps departed Vought in Dallas, destined for Portugal. The planes were manned by a mix of Vought and PoAF pilots and arrived at Monte Real Air Base on the twenty-fourth. They would serve in 302 Squadron. *Author's collection*

The very first TA-7P (#5545, BuNo 153201) prepares to depart Vought for Monte Real in the fall of 1984. Six TA-7Ps would be built, three each going to the two squadrons. Since retirement in spring 1995, it has been fully restored and is on display at the Ovar Air Museum and Sintra Air Base. *Author's collection*

A-7P #5502 (originally A-7A BuNo 153200) banks over the Portuguese coastline. Note the addition of the large UHF dorsal antenna, a feature of both the A-7P and TA-7P. Upon retirement, this aircraft was donated to the Polish Air Museum in Krakow, where it remains on display. *Author's collection*

A four-ship of RTN A-7s, two Echoes and two Twosairs, flies over the iconic Pattaya Park Tower, roughly 20 miles northeast of U-Tapao Air Base. The Thais were proud of their A-7s; sadly, they could not, in the end, afford to maintain and operate them. *Photo by James O'Hora*

The Royal Thai Navy purchased fourteen A-7Es and four TA-7Cs. These airframes were restored by the Naval Air Rework Facility (NARF) in Jacksonville and then delivered to Thailand. TA-7C BuNo 156794 is seen here shortly after departing Jax on its six-day ferry to U-Tapao Air Base, home of the 104th Squadron, who would operate her. *Photo by James O'Hora*

A three-ship of A-7Ps cruise over the Atlantic, date unknown. *Author's collection*

By 2007, all Thai A-7s were grounded. While most were regularly started and occasionally taxied, they would never take to the skies again. Here A-7E 1411 (BuNo 160583) sits in its shelter. Given its tail-high stance, it would appear that its engine has been removed, and the cockpit shows no sign of an ejection seat. *Author's collection*

A look of what might have been a regular occurrence, A-7s flying over the Swiss Alps. While the A-7 performed well, in the end the Swiss rejected all of the competitors, favoring instead to purchase thirty rebuilt Hawker Hunters. *Author's collection*

The USAF loaned Vought two A-7Ds to assist in their efforts to earn a contract from the Swiss air force. In 1972, the Swiss conducted a series of trials and evaluations of the A-7 and its competitors. Here Vought technicians prepare to load 20 mm ammunition into A-7D s/n 71-0294 at Payerne Air Base. *Author's collection*

Rain was a frequent visitor to U-Tapao, as were US contract pilots and maintainers who instructed Royal Thai Navy personnel on all aspects of A-7 operations. *Photo by James O'Hora*

In 1972, a contingent from Pakistan visited the Vought factory to evaluate the A-7 for the Pakistani air force. For this visit, Vought affixed PAF markings on A-7H BuNo 159926. While both Vought and the Pakistanis were very interested in closing the deal, the US government felt the A-7 represented too much of a destabilizing force in the region, and eventually blocked it. *Author's collection*

CHAPTER 8

Twosairs

Even before the first A-7A took to the skies, management at Vought was pitching the idea of a two-seat A-7 to the Navy. Original attempts at wooing the service on the idea of a "Twosair" centered on the concept of a "Super A-7," a supersonic variant, with both upgraded single-engine and twin-engine options proffered. Designed to fulfill fighter, attack, and even reconnaissance needs, conceptual designs for this Super A-7 bore an uncanny resemblance to Vought's XF8U-3 "Crusader III," the company's submission into the competition eventually won by the McDonnell Douglas F-4 Phantom II.

The Navy's reaction to Vought's sales pitch was lukewarm at best. They simply did not see the need for the aircraft and preferred that Vought stay focused on delivering good on their VAL winner. But Vought management remained convinced of the merits of a two-seat A-7, albeit their conviction was based on keeping their production line busy. Still, management at the company was always looking for an opportunity to present the merits of a two-seater to the Navy, and, eventually, other services.

In the spring of 1972, Sol Love, then Vought's A-7 program director, convinced his bosses of the merit in having the company invest in building a two-seat A-7. Love felt that its serving as a proof-of-concept aircraft would finally open the Navy's eyes on the value of such a variant. With the idea green-lighted internally, Love approached the Navy, who, always open to a good deal, offered to bail an A-7E back to Vought so that they could make the conversion. In the end, the Navy directed VX-5 (the test-and-evaluation squadron based at China Lake) to release A-7E BuNo 156801 to Vought to serve as the project—known as V-519—testb ed.

Vought's design required a 16-inch extension to the forward fuselage to provide the needed space for the second cockpit, as well as an 18-inch fuselage "plug" aft of the wing box. The M61 cannon was removed, as were "unnecessary" items such as the ECM equipment. As long as the pilot didn't look backward, his flying experience would be identical to that of a single-seat A-7. The rear/instructor's seat was raised so as to provide better forward visibility, and the rear cockpit included all necessary flight controls and accessed HUD imagery from the front cockpit via a small TV screen, located on the main panel.

The resulting test aircraft was originally designated as the YA-7H, but since it was determined that the H suffix was to be utilized for the Greek (Hellenic) A-7 program, the designation was changed to YA-7E. Regardless of its official designation, this original A-7 two-seater would be best known as the "White Whale," a nickname it acquired due both to the shape of the revised fuselage and the fact that it left the Vought paint shed in an all-white paint scheme. On August 29, 1972, the company's chief test pilot, John Konrad, took the White Whale on its inaugural flight.

Having a flyable demonstrator in the White Whale, and (as A-7E production continued to ramp up) literally hundreds of Bravo and Charlie airframes soon to be retired, the Navy finally saw the need—and the value—in converting some A-7s to two-seaters for use in the RAGs. In the end, the Navy contracted with Vought to produce sixty such aircraft, designated now as TA-7Cs, with twenty-four being produced from A-7B airframes, and the balance of thirty-six from A-7Cs. Like the original, single-seat Charlies, the TA-7C would feature the A-7E avionics suite and a TF-30-P-408 engine. By 1985, the Navy would opt to upgrade forty-nine of their "T-birds" with TF-41 engines and other improvements common to fleet A-7s, such as automatic maneuvering flaps (AMF) and a Stencil ejection seat, which had replaced the venerable Escapac seat.

The White Whale proved to be a great investment for Vought, serving not only as an obvious sales demonstrator for other two-seat A-7 purchases but as a testb ed for equipment and other improvements destined for single-seat A-7s.

In addition to the TA-7C, the two-seat A-7 variants were as follows:

TA-7H

Initial cadres of Greek pilots and maintenance personnel received their training at VA-174, the Navy's East Coast RAG. Impressed with what they saw in the newly arrived TA-7Cs, the Greeks soon added a request for five two-seat variants of the A-7H that were being produced for them. Similar in all ways to the single-seat H models, the TA-7H likewise did not have in-flight refueling capability. In the early 1990s, as refurbished A-7Es were delivered to Greece to replace their well-worn A-7Hs, TA-7Cs would also be delivered.

TA-7P

In 1983, the Portuguese air force (PoAF) placed an order for an additional twenty-four A-7Ps to complement the twenty they had ordered in 1980. Included in this second order was a request for six TA-7Ps. Like their single-seat A-7P counterparts, all six TA-7Ps began their lives as Alphas. Vought would convert these airframes, upgrading them with an A-7E avionics suite. The two-seaters likewise retained the TF-30-P-408 powerp lant used in the single-seat P variant. But while the A-7P kept the twin Mk. 12 20 mm cannons, the two-seaters bound for Portugal did not have a gun.

To familiarize PoAF pilots with their under-construction Twosairs, the US Navy leased them one of their TA-7Cs—BuNo 154404—which served with the PAF until 1985. While in service, it earned the moniker "*Pomba Branca*," or White Dove.

A-7K

By the time the Navy embraced the idea of a two-seat A-7, the Air Force was already starting to withdraw the A-7D from service, preferring to transfer them for use within the Air National Guard. That said, the service did see merit in a two-seat trainer. The first A-7K was built from a converted A-7D (s/n 73-1008), and the remaining thirty that were produced were entirely new airframes. Each Guard A-7D squadron would receive one A-7K for pilot-training purposes; however, they were more frequently used in recruiting/retention, providing incentive rides to enlisted personnel who had reupped their service commitments. The bulk of the A-7K fleet would serve in the 162nd TFW, the Guard's A-7 training wing based in Tucson, Arizona. All K variants would be delivered to the Guard between 1981 and 1983, and most served less than a decade before they were retired.

EA-7L

Modified from TA-7C airframes, the "Lima" served in VAQ-33 (based at NAS Key West) and VAQ-34 (at Pt. Mugu), providing these electronic-warfare squadrons with a platform that could emulate both multiple enemy ECM threats as well as cruise missiles. In the end, the Navy modified ten airframes to EA-7L configuration. The conversion to Lima standards took place in 1983, and like the A-7K, most would serve no more than ten years before retirement.

Only ten TA-7Cs were converted to EA-7L specifications, and pilots of them referred to themselves as "Lima Drivers." *Photo by Lori Gattuso*

Vought was contemplating a two-seat A-7 even before the A-7A prototype had been completed. Seen here are plans for a two-seat, twin-engine variant of the "Super A-7." The resemblance to the company's Crusader III was uncanny. *Author's collection*

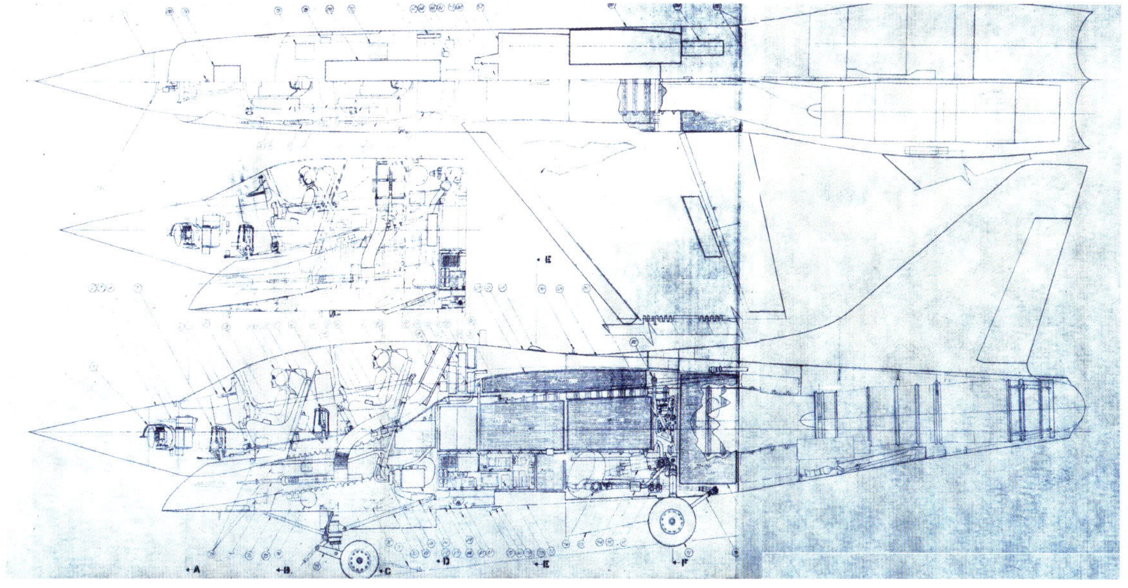

An early Vought mockup of a two-seat A-7E provided designers with a means to determine best cockpit layout. In the end, the front-seat occupant experienced the same conditions as he would in a single-seat—until he looked behind him. *Author's collection*

Most production TA-7Cs utilized A-7B airframes that were lengthened to accommodate a second seat. This was done with the introduction of spliced cockpit area, and a "plug" just aft of the wing box. That plug is easily seen in this photo of BuNo 154361 on the Vought factory floor, circa 1975. *Author's collection*

The YA-7H rolls out of the paint shop, providing a good reason as to why it was called "the White Whale." Note that the aircraft is still stenciled as A-7E BuNo 156801. *Author's collection*

The YA-7H heads out on its maiden flight, August 29, 1972. As with most first flights of A-7 variants, Vought's chief test pilot, John Konrad, is the pilot. The YA-7H was such an amalgamation of off-the-shelf and custom parts that it required its own custom flight and tech manuals. *Author's collection*

The White Whale sits in the fuel pits alongside another Vought one of a kind, the TF-8A. The original design for a two-seat A-7 called for a canopy like the TF-8's, but it was later determined that boarding the aircraft would be easier with a canopy hinged on the starboard side. Note the addition of a gold band on the tail, surrounding the original Corsair II Squared logo. *Author's collection*

Soon after its initial test flights, the White Whale hit the road in an effort to drum up sales. Seen here in custom VA-174 livery, it flies lead on a "Hell Razor" Echo (A-7E BuNo 158763) as they approach NAS Cecil Field. *Author's collection*

The very first TA-7C is delivered to VA-122, May 23,1978. VA-174 would start getting theirs the following month. Originally starting life as a Bravo (BuNo 154379), it began its career ten years earlier, serving in VA-122. It would spend another ten years with VA-122, ending its career in VX-5 and then NATC. *Author's collection*

Wearing a new tail art, the White Whale sits on the Vought ramp. Note that the ECM fairing now sports an anticollision taillight where the antennas used to be. This view also provides a good look at the tailc one faring, which contained a spin parachute. While this faring was built into production TA-7Cs, the spin parachute was never used in the RAG. *Author's collection*

Starting life as an A-7B, BuNo 154477 would be the first airframe converted into a true TA-7C. As such, it spent most of its life assigned to NATC, performing the bulk of testing on the type. This photo provides another good look at the spin chute faring. *Author's collection*

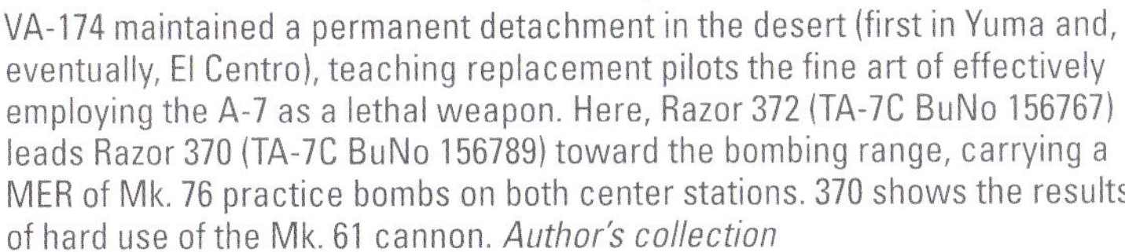

VA-174 maintained a permanent detachment in the desert (first in Yuma and, eventually, El Centro), teaching replacement pilots the fine art of effectively employing the A-7 as a lethal weapon. Here, Razor 372 (TA-7C BuNo 156767) leads Razor 370 (TA-7C BuNo 156789) toward the bombing range, carrying a MER of Mk. 76 practice bombs on both center stations. 370 shows the results of hard use of the Mk. 61 cannon. *Author's collection*

Eagle 221 (TA-7C BuNo 154402) flies wing on Eagle 201 (A-7E BuNo 160858) somewhere over Fallon, circa 1980. During this timef rame, VA-122 aircraft were stenciled with "USS Lexington." While the squadron CQs often used the *Lex*, they were not assigned to it. TA-7Cs were fully capable of carrier operations but seldom seen on the boat. In fact, by the mid-1980s, RAG T-birds would have their launch bars removed. *Author's collection*

The Naval Weapons Center in China Lake was tasked with developing the FLIR pod for fleet use. The project resulted in the "Night Raider" emblem affixed to the tail of test aircraft. Single-seat A-7s would be similarly painted, but with just one pirate instead of the two seen here on BuNo 156768, photographed at China Lake in June 1980. *Photo by Michael Grove, Naval Aviation Museum*

TA-7Cs were employed at the Pacific Missile Test Center, serving in roles from chasing launched cruise missiles to, as seen here, ECM threat simulators. Bloodhound 81 (TA-7C BuNo 154464) carries the ALQ-167 tactical contingency pod on station #1. *Date and photographer unknown, Naval Aviation Museum*

Five TA-7Hs were built under contract for the Hellenic air force. The first, BuNo 161218, had its maiden flight on March 4, 1980. All five TA-7Hs would fly with the HAF for over twenty-five years. *Author's collection*

The first A-7K was a converted A-7D, s/n 73-1008. Thirty A-7Ks were to follow, each being built from the ground up. Seen here on the Vought ramp shortly after conversion, the prototype A-7K is still awaiting installation of its ejection seats. It would first fly on November 7, 1980. *Author's collection*

While the Portuguese air force waited for delivery of their TA-7Ps, the Navy leased them TA-7C BuNo 154404. The PoAF pilots referred to it as "*Pomba Branca*," or "White Dove," It would serve in the PAF for three years and ended up its service life with the Hellenic air force. *Author's collection*

Starting in 1983, ten TA-7Cs were converted to EA-7L configuration, serving with two electronic-warfare squadrons. VAQ-34 was the West Coast recipient of the Lima, flying out of Pt. Mugu. Here Flashback 205 (EA-7L BuNo 156761) carries an AST-4 threat simulator pod on station #2. *Naval Aviation Museum*

The only airworthy A-7 to receive a FAA civil N-number was this TA-7C (BuNo 154477). In August 1993, the Navy registered it on the behalf of Thunderbird Aviation, a Phoenix-based defense contractor. Within three years it would return to storage in Tucson, and it was sold to the Greeks in 2000, where it would serve until the type's retirement in 2014, having the distinction of being the last A-7 to take to the skies. *Naval Aviation Museum*

CHAPTER 9

The YA-7F

Given the Air Force's long-standing penchant for speed, it was puzzling to many that the A-7D had been replaced by the Fairchild-Republic A-10 Thunderbolt II. Compared to the A-10, the A-7 was practically a hot rod—faster and at least as (some would say more) maneuverable than the aircraft that replaced it. What the A-10 *did* have going for it was that it was an all–Air Force program. By 1985, the Air Force was concerned enough about the A-10's performance that it issued an RFI (request for information) to contractors, asking for them to submit reports on the options for aircraft that could perform the missions of close air support and battlefield interdiction better than what they currently had. Vought responded, but not just with a report. Instead, they presented an all-encompassing effort—an effort that proposed that the Air Force commit to an upgraded A-7D.

Vought's belief in the "Super A-7" concept went back to at least 1965. Having twenty years' worth of research, hundreds of A-7Ds currently serving with Air National Guard units across the US and Puerto Rico, and (admittedly) an urgent need to keep the production line open, the company was convinced that they could make a compelling argument. In the end, Vought's persistence earned it a contract to develop two prototypes, designated the YA-7F. Over the course of its life, the aircraft would be known alternatively as either the "A-7 Plus" or (more frequently) "Strikefighter."

In creating the YA-7F, Vought finally had an aircraft that not only bore a superficial resemblance to its F-8 Crusader. It truly was an amalgamation of the best attributes of both the F-8 and A-7. It would be powered by a Pratt & Whitney F100-PW-220 with afterburner (the same powerp lant used in the F-15 Eagle), which would provide 73 percent more thrust than the TF-41-powered A-7D. To accommodate the larger engine, the fuselage was extended by 4 feet, with plugs both forward and aft of the wing box. An additional 10 inches was added to the vertical tail, with the horizontal tails incorporating a marked anhedral, as opposed to the dihedral that other A-7 variants sported. Additional improvements included updated avionics, pilot oxygen system, and night attack capabilities. If the test program proved successful, Vought proposed rebuilding 307 A-7Ds to A-7F standards and stated they could do so at a unit flyaway cost of just under $5 million—a relative bargain.

Testing of the YA-7H prototypes was conducted at Edwards AFB, with Vought's chief pilot, Jim Read, heading up the flight test program. Read, along with Vought's Jerry Mumfrey, would perform the bulk of the flying. The first YA-7F (s/n 71-0344) took to skies on November 29, 1989, with Read at the controls. The flight was just over one hour in length and proved satisfactory enough to permit it to go supersonic on the following flight. The second prototype (s/n 70-1039) would make its first flight on March 3, 1990. Over the course of the nine-month test program, both prototypes were put through their paces and proved themselves to be strong performers.

In the end it was not to be. The Air Force passed on the A-7F outright. Officially, the Air Force would state that it did so due to a convergence of budget constraints and the fact that its primary adversary—the Soviet Union—was in its death throes (it would dissolve at the end of 1991), negating the need for the aircraft. In reality, the A-7F program was considered by most insiders to be "dead on arrival." The Air National Guard was clearly committing itself to the F-16, with Virginia's 192nd TFG slated to retire their A-7Ds in 1991, becoming the first Guard unit to deploy the Fighting Falcon. The writing was on the wall for all to see.

Without a contract for the A-7F from the Air Force, Vought was facing truly bleak times. The last A-7K had rolled off the assembly line in 1984, having the dubious honor of being the last all-new aircraft the company would ever build. Vought tried to sell the concept of the Strikefighter on the export market but found no takers.

Even to this day there are those who feel the Air Force had once again given short shrift to the A-7, favoring instead an all-new, all-USAF program. Clearly, the A-7F was equal in performance to the F-16 on the battlefield. It could carry more ordnance, had a longer combat radius, and could be acquired for a fraction of the cost of a new F-16. At least on paper, the Strikefighter was a winner. In reality it never stood a chance.

Vought conceptual artwork of the proposed A-7F. *Author's collection*

Two A-7D airframes were bailed out to Vought for conversion into YA-7F test aircraft. Undergoing conversion is s/n 70-1039 (foreground), which would become the second prototype, and s/n 71-0344 (*background*), which would become YA-7F #1. *Author's collection*

YA-7F #1 (s/n 71-0344) is rolled out from the hangar, early in the morning of its official debut to the press. Note that the A-7's signature notched fin cap has been replaced by a straight one. *Author's collection*

The YA-7F would be powered by a Pratt & Whitney F-100-PW-220 engine with afterburner. Capable of 26,000 pounds of thrust, the YA-7F could sustain Mach 1.5 in level flight. *Author's collection*

Vought's chief test pilot, Jim Read, prepares for the first flight of the YA-7F, which took place on November 29, 1989. This kicked off a nine-month test program that put both prototypes through their paces. Note the larger, single pitot tube, which replaced the twin pitot tubes of earlier A-7s. Both test articles initially featured an additional probe mounted on the radome. *Author's collection*

Read shared test flight responsibilities with Vought's Jerry Mumfrey, seen here departing Edwards in full burner in prototype #2. The second YA-7F first flew on April 3, 1990. Note that the horizontal tail has noticeable anhedral to it, whereas all other A-7 variants had significant dihedral to them. *Author's collection*

Prototype #1 makes a simulated emergency landing at Edwards. Notice the EPP (emergency power package) is deployed. This view also shows off the wing leading-edge extension, and the ventral strakes on either side of the tailhook. *Author's collection*

Early on in the testing, the YA-7F was certified for in-air refueling, with the first such test being made in this photo. *Author's collection*

Prototype #1 flies formation on A-7D s/n 69-6195, the twenty-fifth A-7D built. The photo provides a nice comparison between the two variants. *Author's collection*

CHAPTER 10

Going Out with a Bang

In September 1984, the Air National Guard took delivery of A-7K s/n 81-0077, the thirty-first K model built. Its delivery marked the end of the production of all A-7 aircraft, a production run that saw 1,545 airframes produced over a span of nearly twenty years. The delivery also marked another, sadder milestone in that 81-0077 represented the very last aircraft to be built by the Vought Corporation. While the company would soldier on (and continues to do so, in name at least) in the production of subassemblies for other aircraft manufacturers, there would never be another aircraft built that carried a Vought data plate.

With the introduction of the F/A-18 Hornet, A-7E squadrons were slated either for transitioning or decommissioning. In May 1983, VA-25 became the first fleet squadron to transition to the Hornet, reporting to VFA-125, the former A-7A/B RAG, which was now training Hornet pilots and maintainers. VA-174, the East Coast A-7 RAG and once the largest squadron in the entire Navy, was decommissioned on June 30, 1988. VA-122, its West Coast equivalent, would follow suit on May 31, 1991. The squadron would be recommissioned in January 1999 as VFA-122, the West Coast Super Hornet RAG. By the end of 1990, all but two fleet Echo squadrons had gone out of business. And those two—VA-46 and VA-72—had been in the midst of transitioning to the Hornet; that was, until Saddam Hussein's invasion of Kuwait changed those plans.

Having returned to Cecil Field in June 1990 from a short cruise on *John F. Kennedy*, the "Clansmen" of VA-46 and the "Bluehawks" of VA-72 started cycling pilots and maintainers through VFA-106, the Hornet RAG, located just down the ramp. A number of their newer Echoes had already been transferred to VA-37 and VA-105. On August 2, the Iraqi army had invaded Kuwait, and the following day skippers of both transitioning squadrons were discussing with CAG-3 what it would take to restore them both to ready status with their A-7s. Within a week, *Kennedy* and CAG-3 were on the move, and VA-46's and VA-72's Echoes with right there with them.

At the time, the A-7E was equipped with a better HARM system than that employed in the Hornet. As a result, 46 and 72 would primarily be tasked with the mission of taking out enemy SAM installations: Ironhand, a job the A-7 proved so excellent at over the skies of Vietnam some twenty-four years earlier. Operation Desert Storm was launched on January 17, 1991, and the Echo squadrons acquitted themselves exceptionally well, knocking out SAM sites and delivering CBUs as well as both iron and smart bombs throughout the duration of the campaign. Neither squadron would suffer a combat loss; however, VA-72 *did* lose an aircraft on January 24, due to nose gear failure on launch. Both plane and pilot recovered aboard *Kennedy* with the use of the barricade.

Kennedy and CAG-3 returned home on March 28, 1991. The "Clansmen" and "Bluehawks" would retire their A-7Es by June, with both squadrons being decommissioned that same month. It was only fitting that the aircraft thrown into the heat of battle two years after its maiden flight would retire from frontline service having proved itself again to be a capable warrior twenty-four years later.

The year after Desert Storm would see the retirement of most Air National Guard A-7Ds and Ks, with the last squadrons transitioning to other platforms in 1993. Tulsa-based 125th TFS traded in their Big Ds for F-16s in October 1993, bringing an end to the A-7's USAF/ANG chapter. October 1993 would also see the Navy's last squadron, the "Flashbacks" of VAQ-34, retire their EA-7Ls. Come November 1993, if you wanted to see an A-7 take to the skies, you would have to travel outside the United States to do so.

In March 1993, the Hellenic air force took initial deliveries of A-7E and TA-7C aircraft. In the end, forty-two Echoes and eighteen

Twosairs flew with the HAF. All were former US Navy aircraft, many of them veterans of Desert Storm. Like the H models before them, the Greeks put these sixty aircraft to good use, flying them for two more decades before it was time to bid them adieu. By November 2012, only one squadron in the world was still flying A-7s. That honor went to the HAF's 336 Squadron—the Olympos squadron, based at Araxos. On October 17, 2014, the squadron hosted the official retirement ceremony for the A-7, bringing together Corsair II veterans from around the world to pay their respects.

One week after the official retirement ceremony was held, 336 Squadron held another, private ceremony of their own. On October 24, squadron commander Apostalos "Papilos" Papadopoulos climbed into TA-7C, BuNo 154477, and in the presence of his fellow Olympos pilots and maintainers fired up its TF-41 and took her aloft one last time. Since Papilos's flight, the sights and sounds of airborne A-7s are no more, save for in the memories of those who fondly recall its decades of faithful service. While often overlooked by those who see beauty as only skin deep, the A-7 Corsair II was truly an amazing aircraft. It ushered in a new age of technology, and with it, a versatility, economy, and capability not to be found in other military aircraft of its day. For those reasons alone, the A-7 deserves to be regarded as one of the best bang-for-the-buck warplanes to ever take to the skies.

Vought employee Brett Buie pulls A-7K s/n 81-0077 out of Building #6, heading to the paint shed. It was the thirty-first A-7K built and was the end of the line. It was not only the last A-7 produced, but it was the last complete aircraft that Vought ever produced. Before it was painted, Vought employees signed their names to its skin. *Author's collection*

The "Fist of the Fleet" became the first A-7E squadron to transition to the F/A-18 Hornet, with ceremonies taking place in May 1983. In replacing their Echoes, squadron personnel reported to VFA-125, the Hornet RAG that once served as the A-7A/B RAG. *Author's collection*

VA-46 was in the midst of transitioning to the F/A-18 when they got the call to muster their men and A-7s. This patch was worn during the short-lived transition. *Author's collection*

VA-72 Echoes lined up on the bow of *Kennedy*, preparing for a mission. The pilots have manned up, and most Echoes are loaded up with Rockeye CBU canisters. The day after Kuwait was invaded, both they and sister squadron VA-46 reassembled their A-7s and the necessary personnel and one week later were heading into action. *Author's collection*

Decoy 403 (A-7E BuNo 159999) loaded with eight Mk. 82 Snakeyes. To improve performance of their aircraft, Desert Storm A-7s had one wing station removed. Not all configurations were symmetrical, as evidenced in this photo, where stations #3 and #7 have been removed. This was done to provide as much flexibility of load-out options. *Photo by John Leenhouts*

USAF KC-135s, a.k.a. "Iron Maiden," provided much-needed refueling services for Navy aircraft during Operations Desert Shield and Desert Storm. An A-6E of VA-85 hits the tanker over Saudi Arabia as VA-46 Echoes wait their turn. *Photo by John Leenhouts*

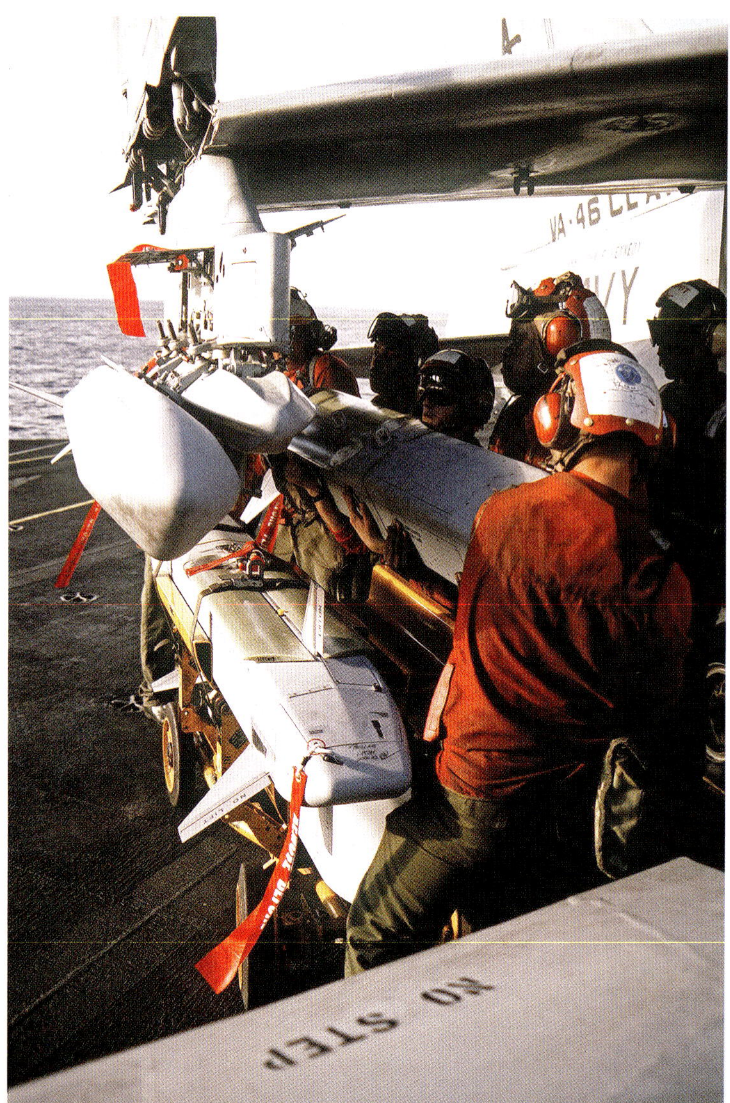

Ordies from VA-46 load TALD decoys on Lt. Jeff Greer's Echo on January 16, in preparation for the first mission of Desert Storm. Greer would enter Iraqi airspace ahead of his squadron mates, launching his decoys in an effort to light up enemy SAM sites. Other "Clansmen" aircraft, loaded with HARM missiles, would then take out those sites. *Photo courtesy of Jeff Greer*

Night one operations on *Kennedy,* a VA-46 Echo on cat #3, with a VA-72 bird on #4, preparing to launch, while other A-7s queue up for their turn. *Photo courtesy of Jeff Greer*

Practice missions over northern Saudi Arabia proved vital to the success of Operation Desert Storm. Here a VA-72 A-7 prepares to enter the region's "Star Wars canyon." *Photo by John Leenhouts*

At the time of Desert Storm, the A-7E was a better HARM delivery platform than the Hornet, making it a vital component to the war plans. *Author's collection*

VA-72 would lose one A-7 during Desert Storm, but not to combat. On January 24, 1991, Decoy 402 (A-7E BuNo 158830) suffered nose gear failure on launch. It was brought back aboard *Kennedy* by means of the barricade. Given the aircraft would soon be retired, it was decided that it was better to strip it of all useable items and give it a burial at sea. Here it is seen going over the side on January 25. *Author's collection*

Kennedy and CVW-3 returned home to a hero's welcome on March 28, 1991. VA-46 and VA-72 were disestablished on June 30. Decoy 400 (A-7E BuNo 160552) is seen here on the ramp at Cecil Field, May 2, wearing a Desert Storm commemorative paint job. It would soon head to the rework facilities at NAS Jacksonville in preparations for transfer to the Greeks, where it would serve until retirement of the type. *Author's collection*

What was originally described as a "three-month saber-rattling cruise to the Red Sea" turned out to be a seven-month combat cruise. With Operation Desert Storm a success, the USS *Kennedy* and Air Wing 3 headed for home. Here VA-46's XO, J. R. Stevenson prepares to launch from the *Kennedy* for the last time. March 27, 1991. *Photo courtesy of J. R. Stevenson*

On April 5, 1991, VA-122 held a ceremony at NAS Lemoore to officially deactivate the squadron. Squadron instructors served as "pall bearers," burying a squadron time capsule on NAS grounds. The squadron would be reactivated in 1999 to serve as the West Coast Super Hornet squadron. *Photo by David F. Brown*

The past flies with the future. VA-122 deactivation ceremony, April 5, 1991. *Photo courtesy of Jeffery Harrison*

Part of the A-7 retirement ceremonies held at Araxos Air Base, October 17, 2014. *Author's collection*

336 Squadron commander, Apostalos "Papilos" Papadopoulos (*left*), and Petros Hatziris (*right*), commander of the 116 Combat Wing, prepare to board in TA-7C BuNo 154477 for the final flight, October 24, 2014. *Photo courtesy of Papilos Papadopoulos*

336 Squadron's commander, Apostalos "Papilos" Papadopoulos, had the honor of piloting the last-ever A-7 flight. Held a week after the official retirement ceremony, this flight was a private affair, made in the presence of squadron personnel. *Photo courtesy of Papilos Papadopoulos*

The A-7 Corsair II ushered a bevy of new technologies, resulting in an unheard-of level of accuracy *and* economy. Beloved by those who built, maintained, and flew her, the A-7 has earned its place as the best bang-for-the-buck attack aircraft ever produced. *Author's collection*